For a complete listing of titles in the
Artech House Mobile Communications Library,
turn to the back of this book.

Signal Failure

The Rise and Fall of the Telecoms Industry

John Polden

ARTECH
HOUSE

BOSTON | LONDON
artechhouse.com

Library of Congress Cataloging-in-Publication Data
A catalog record for this book is available from the U.S. Library of Congress.

British Library Cataloguing in Publication Data
A catalogue record for this book is available from the British Library.

Cover design by Joi Garron

ISBN 13: 978-1-68569-057-1

Contents

Preface

This book is a result of a personal journey for me. Having spent my career involved with U.K. technology businesses, I wanted to step back and look in more detail at how technology affects business and society. I have always been interested in the history of enterprises and I noticed how quickly once-thriving businesses could disappear with little trace. Unlike most major industries of the past (steel, shipbuilding, cotton), little physical evidence of modern technology businesses survives when they cease trading.

Once I had decided that I wanted to research this area, I asked around as how to best approach this task. I was fortunate to meet up with Sir Geoffrey Owen, who, among many things, had written books on British industry similar to my area of interest. In turn, he introduced me to Professor Ed Steinmueller of the Science Policy Research Unit (SPRU) at the University of Sussex, who had extensively researched and written on this area. As a result, I was fortunate to become an Honorary Research Fellow at SPRU.

Looking back, I am amazed to find that I have spent 7 years at SPRU. Perhaps I am a slow learner, but I had to learn the basics of academic research and writing under the patient guidance of Professor Steinmueller. I discovered there was a big difference between my world of business and the one of academic research that I have attempted to bridge. Any weaknesses in academic research are entirely down to me.

As an electronics engineer, I decided to particularly look at how the invention and development of semiconductors had affected key industries in the period leading up to 2000. I came to realize that there was a largely untapped resource of recent historical information available to me from people who had lived through the period that I chose to cover. These people were helpful and encouraging for my project and I will be forever grateful to them.

I had initially aimed to cover the entire electronics industry but came to the decision that it would be best to initially concentrate on one sector of that industry and focus on the United Kingdom. The choice of telecoms was largely because I could see how it spawned innovations (such as social media) that were inconceivable to the developers of the key technologies on which it was based. Hence, this book looks at how the United Kingdom fared with the development of the modern telecoms industry. Here I could explore the received wisdom that the United Kingdom is good at inventing but weak at commercialization.

Having decided to aim my research at producing a book, I must thank all those people who helped me in this enterprise. In particular, Professor Steinmueller gave me extensive feedback on the numerous drafts of my book. I also received valuable help from Visiting Professor Andy Valdar of University College London on the technical issues of telecoms with which I had to wrestle. Finally, I would like to thank Fred Goldstein of Ionary Consulting, who reviewed my draft on behalf of the publishers and provided me with very helpful input, particularly relating to the U.S. telecoms industry.

However, in the time-honored practice, my greatest thanks go to my wife, Helen, who indulged my need to sit in front of a computer for hours when there were so many other things we could do in our busy retirement.

1

The World of Telecoms

1.1   Communications Count

In January 2020, the media was full of a story about how the United Kingdom's national security is threatened by plans to install fifth generation (5G) mobile telephony supplied by Huawei, a Chinese supplier closely linked to the Chinese government. Somehow the Chinese would get to know every conversation and communication in the country, even the contents of everyone's freezer.

The outcry led to rapid backtracking by the U.K. government. In July 2020, the government stated that Huawei equipment already installed would be removed from strategically sensitive areas of the communications network. Alternative suppliers would be used from "friendly" countries.

Throughout the episode, all commentators accepted that the communications network of the country was a key strategic asset. The onset of 5G (see Chapter 3) was seen as a major step in the increasing interconnectivity of the country. However, few commentators paused to consider how the United Kingdom, with the world's fifth biggest economy and a long proud history of technological innovation had got itself into a situation where it seemed to be totally dependent on foreign suppliers for this key part of its infrastructure.

In this book, I seek to redress this balance. It looks at telecommunications as it developed throughout the world and then focuses on what happened in the United Kingdom in the crucial period of 1950 to 2000 when the forces of technological change, economic growth, and globalization of supply combined to create the situation in which we are now. Today the U.K. telecoms supply industry is pretty much totally dependent on imported products with no major U.K.-centered supplier to address the country's communication needs.

At this point, I should point out that what is widely termed the telecoms industry is usually split by analysts into two major distinct components. The telecoms supply industry develops, manufactures, and installs communications equipment. Telecoms operators purchase the equipment and operate it. The operators (such as British Telecoms (BT) and Vodafone) are much more widely known as they are the people who send us our telephone bills. However, I believe that it is the telecoms supply industry that is strategically important as they supply the means by which telecommunications is developed and installed.

I will look at how the U.K. telecoms industry responded to changes in the market for telecommunications equipment, particularly from 1950 to 2000. During this time, there was a massive increase in demand for telecoms services due to growing wealth, technological changes, and market deregulation. These changes required the industry to adapt rapidly. I shall compare how the U.K. industry fared compared with other appropriate countries and lessons taken that are potentially applicable to other industries. In these 50 years, the United Kingdom went from being one of the biggest exporters of telecoms equipment to being one of the biggest net importers (see Appendix A).

However, I seek to go beyond documenting the business and economic history of an industry sector. There are important issues that need to be addressed. Why did the United Kingdom, a leading technological power, fail to develop a telecoms supply industry as a means of economic growth? Why was the strategic importance of communications not recognized (or at least acted upon) by successive U.K. governments? Do we observe similar failings in other industries? Can we expect similar issues in other key technology industries? Before I attempt to answer these questions, I lay out the background and history of this industry.

I believe that there are lessons that are highly relevant today well beyond the telecoms industry. The issues of a post-industrial Britain have continued to exercise commentators. Some have observed the rapid decline of traditional industries particularly as the country adapted to joining the Common Market (as it then was) in 1973 [1]. Others see it more in terms of the decline of a once-great nation under the guidance of incompetent politicians and institutions [2] often highlighting the demise of once-world-leading industries such as motorbikes [3]. While manufacturing has been a declining part of the U.K. economy for most of the twentieth and twenty-first centuries, having been trained as an engineer, I believe that having a strong manufacturing base has an importance beyond its proportion of the gross domestic product (GDP). This issue has been widely explored for many post-industrial economies, notably in the United States. Some commentators are concerned that a declining manufacturing base [4] weakens the economic viability of a country, although other commentators believe that, in this era of knowledge-based economies, the physical manufacturing of goods is best left to lower-cost producer countries, even if it does create a large and consistent hole in the finances of such countries [5]. I leave it to the reader to decide whether they believe that a strong manufacturing base is important to the future of the United Kingdom. Certainly, if the United Kingdom does not believe it needs a strong manufacturing base, it has been successful in following through on this policy. Using the telecoms industry as an example, I will explore how the United Kingdom performed in this rapidly growing technology-based industry and the factors that drove the outcome during the period.

In developing this history, it became clear that there are messages that are relevant beyond the U.K. telecoms industry. Some of the conclusions that I draw relate to the economic, political, and cultural systems of the United Kingdom. These factors had similar effects on other key U.K. industries and will continue to do so when new industries of the future such as biotechnology evolve in the United Kingdom.

1.2 The United Kingdom Then and Now

Let us start by looking at the United Kingdom in 1950 and how it compares with 2000. In later sections, I will expand on how the United Kingdom emerged from the trauma of war and sought to develop new industries (such as electronics) to fill the void left by its long declining industries (such as cotton).

It is tempting to gloss over what a grim time the 1950s were. The black-and-white photography of the period is very apposite, as it was in many ways literally a gray era (such as the famous London smog of 1952). The mainly coal-powered economy struggled to meet the aspiration of the "land fit for heroes" offered by politicians during World War II. The foundations of a welfare state had been largely laid by 1950 but at the cost of depressed economic growth with rationing continuing to 1951.

By 2000, the United Kingdom was a very different place. The adoption of color in the media from the 1960s onwards and the decline of smokestack industries literally made the country brighter. This was echoed in social changes with the prevalence of youth culture and a much more ethnically diverse population. With this came liberalization in key areas such as capital punishment and attitudes toward sexuality. Perhaps the key difference was that everyone was just much richer (about 3 times higher GDP per head). With increasing wealth came other issues such as an increasing concern about substance abuse (alcohol, tobacco, and drugs) and the allocation of resources. However, it is hard to dispute that the United Kingdom in 2000 was a much more pleasant place than in 1950.

In the world of communications, changes were dramatic. In the 1950s, the country was only at the beginnings of the television (TV) era with the state broadcaster, British Broadcasting Corporation (BBC), transmitting to 350,000 TV license holders for black-and-white televisions. To most people, telephony was something used by businesses and the wealthy, with only 1.8 million residential telephone lines serving a population of around 50 million (see Figure 1.1). The iconic red telephone box was the nearest that most came to accessing the telephone service.

In contrast, by 2000, almost all U.K. homes had a telephone line and TV. However, that is not the most remarkable change. The mobile phone, which was almost unknown in 1950, was now part of popular culture, with over one-quarter of the world's population owning one. In the United Kingdom in 2000, the number of mobile phones per head of population was approaching 80% (see Figure 1.2). TV services had mushroomed with many channels and a faint memory of the previous black-and-white era. The development of the internet was under way, opening the door to the coming era of on-demand entertainment and social media. The effects of these changes were not appreciated in 2000 and are still reverberating through society.

In turn, these changes had profound effects on the U.K. telecoms industry. The state-controlled monopoly service provider General Post Office (GPO) changed into a private company, BT, although it maintained a near monopoly on fixed telephony. However, competing service offerings for mobile phones

Figure 1.1   Telecoms 1950s style. (Source: Sahara Prince, Shutterstock ID 165458087.)

Figure 1.2   Telecoms 2000 style. (Source: Akhenaton Images, Shutterstock ID 393607117.)

came from separate independent companies. In line with the globalization of commerce in the period, the ownership of these companies was increasingly from outside the United Kingdom. The trend in globalization also had a critical effect on the telecoms supply industry. Although in 1950 it would be almost unthinkable for the GPO to procure its main equipment from outside the United Kingdom, by 2000, the bulk of telecoms equipment would be imported.

This, in turn, led to the dramatic decline in the U.K. telecoms supply industry and the disappearance of most of the established names and the factories that they operated. Unlike the traditional industries that preceded them (cotton, coal, steel, shipbuilding), their demise left little physical evidence of what were once-significant enterprises. Figures 1.3 to 1.5 show the evolution of the GEC Telephone Works Coventry.

In this book, I map the evolution of telecoms, both from the increasing role it plays in society and how the telecoms industry reacted to these changes. It is the first part of an overall study I am undertaking looking at the U.K. electronics industry and how it performed in providing the economic growth and employment that the country needed. The book focuses on the 50-year period from 1950 to 2000 where I map the size of the industry and compare its performance with that of comparable economies, notably France, Germany, Japan, and the United States. The year 2000 coincides with what was referred to at the time as the dotcom crash [6]. By that point, the electronics industry had, to a considerable extent, become mature with a vast global market with a substantial part of

Figure 1.3   1939. (Source: www.telephoneworks.co.uk.)

Figure 1.4   1990. (Source: www.telephoneworks.co.uk.)

Figure 1.5 2010. (Source: www.telephoneworks.co.uk.)

its hardware imported from Asian manufacturers. Key driving innovations in this period were the development of semiconductors and the resulting evolution of a software industry. These were the innovations that shaped the way the world developed in this period and the consequent growth in wealth. By 2000, a large part of electronics manufacturing had migrated to Asia where low-cost producers such as China and former low-cost producers such as South Korea used the growing demand in electronics goods to dramatically expand their economies. Initially, they concentrated on relatively simple subcontract manufacturing and component supply, but by 2000 had developed sophisticated electronics industry to increasingly challenge established European and American suppliers.

The increasing availability of low-cost platforms for data processing had also spawned the information industry of the internet, social media, and mobile data. As I wrote this book, it became clear to me that there were times where I had to mention events that occurred after 2000. While I maintain the belief that the key actions took place by the end of the millennium, some of the consequential events needed to be recorded. In particular, in the world's telecoms supply industry I was concerned with several key players: notably Lucent (formerly Western Electric) of the United States, Marconi (formerly GEC) of the United Kingdom, and Nortel (formerly Northern Telecoms) of Canada all disappeared as independent businesses in the period up to 2009. However, I have tried to avoid displaying "perfect hindsight" and only mention such events where they are a clear consequence of what took place in the period from 1950 to 2000.

As I assembled material, it became clear that a simple chronological approach would not be appropriate. In 1950, the telecoms industry could be usually defined by one national service provider and a few national equipment suppliers focusing heavily on telephony. However, major technology and market changes greatly broadened the scope of analysis. In particular, the development of mobile telephony and data communications mentioned above had an enormous effect on the world economy and culture. For this reason, I have looked at the development of fixed-line communications (Chapter 2), mobile telephony (Chapter 3), and data communications (Chapter 4) in separate sections. Each of these areas did not develop in isolation but have fed on parallel developments in other areas. The results of this are brought into the effect on the telecoms supply industry in Chapter 5 where its performance is discussed and the reasons behind the supply situation described at the beginning of this chapter are highlighted. The subsequent conclusions follow with suggestions of broader lessons to be learned.

1.3   The U.K. Economy

This section looks at the political and economic background in which the U.K. telecoms industry operated and the major changes that it faced in the period from 1950 to 2000. It was a time when most countries, including Britain, enjoyed a dramatic increase in wealth. The United Kingdom's GDP/head increased in real terms by more than 3 times, according to the Office of National Statistics (ONS). Technological advances were a major factor in the generation of this wealth. However, in 1950, things looked very different than they do now, with the world only just emerging from major conflict and most governments focused on basic issues such as feeding their population. By 2000, these basic issues scarcely merited thought in developed economies and the primary issues were often about excess, social issues, and the consumption and allocation of resources.

1.3.1   The United Kingdom in 1950

By 1950, the postwar chaos had begun to subside, and longer-term issues could begin to be addressed. Britain faced a difficult future. It had won the war, but at an enormous price. It had exhausted most of the wealth that had been accumulated as a leading economy for the previous 200 years. Britain's position as an imperial power was increasingly untenable with independence for India in 1947 and the likely knock-on throughout the huge British Empire. With this would go the advantageous position that British industry enjoyed with respect to trading within its empire and former colonies. Losing this advantage would require industry to compete on a much more equal basis. Britain's traditional industries of textiles (see Figure 1.6), coal, steel, and shipbuilding were particularly exposed to these political changes and would face continuing decline. In many cases, this decline was not new and started much earlier in the twentieth century.

However, there were clear opportunities. British technology had been a major factor in the winning of the war. The country was in a strong position to develop the new industries that the war had stimulated: aerospace, nuclear, computing, electronics, and the life sciences. Could these industries provide enough growth to compensate for the likely decline in Britain's traditional industries? Governments clearly supported this concept and made efforts to encourage these new industries. These efforts would require not just the ability to conceive of new technologies (which was often regarded as a strength of the United Kingdom), but also to develop and market these technologies, as demonstrated in Figure 1.7. Using new technologies as an engine for economic growth ran into issues with the U.K. economy, which we explore later. These go back to Victorian times and relate to how the country adopts and develops new technologies. The United Kingdom's track record here had been mixed with the country slow to develop electricity-based and mass assembly industries. However, with some industries, such as civil engineering, the United Kingdom had maintained a world-leading capability.

Figure 1.6   When cotton was king. (Source: Everett Collection, Shutterstock ID 237232108.)

Figure 1.7   Brabazon, a failed precursor to wide body jets. First flew in 1951; was scrapped in 1953. (Source: BAE Systems.)

1.3.2   U.K. Economic Performance

The history of the U.K. Industrial Revolution has been covered in detail by many economic historians [7]. It might seem surprising that the country largely acknowledged as the first to engage in major industrialization and hence benefit from the consequent industrial revolution should be considered weak at adopting new technology. However, economic historians have pointed out that although the United Kingdom benefited from the growth of a steam-power based economy in the period from 1770 to 1870, it failed to adopt later technologies (such as electricity) as successfully as other nations, notably the United States [8]. The relative performance of the U.K. economy since the beginning of the Industrial Revolution to recent times and, in particular, the productivity of manufacturing has been compared with other economies by analysts such as Broadberry [9]. The resulting analysis has highlighted that the U.K. economy has always struggled to achieve consistent high economic rates of growth. This is shown in Tables 1.1 and 1.2, which compare the United Kingdom to Germany and the United States.

These are derived from the work of an economist, Andrew Maddison, who looked at the GDPs of countries throughout history. It shows some surprising results if you look at the three major economies shown [10]. The numbers shown for Germany in 1950 to 1989 are solely for West Germany and hence somewhat distort the picture.

The United Kingdom had a significantly higher GDP per person in 1700 before the industrial revolution. Even in its period of economic leadership

Table 1.1
GDP per Person in U.S. Dollars $1,990

	1700	1820	1870	1913	1950	1973	1989	2008
United Kingdom	$1,250	$1,706	$3,190	$4,921	$6,939	$12,025	$16,414	$23,742
Germany	$910	$1,077	$1,839	$3,648	$3,881	$11,966	$16,558	$20,801
United States	$527	$1,257	$2,445	$5,301	$9,561	$16,689	$23,059	$31,178

Table 1.2
Growth Rate per Year

	1700–1820	1820–1870	1870–1913	1913–1950	1950–1973	1973–1989	1989–2008
United Kingdom	0.3%	1.7%	1.3%	1.1%	3.2%	2.3%	2.3%
Germany	0.2%	1.4%	2.3%	0.2%	9.1%	2.4%	1.3%
United States	1.2%	1.9%	2.7%	2.2%	3.2%	2.4%	1.9%

(1770 to 1870), the U.K. economy did not achieve much more than 1% per annum real average growth in GDP/person. From a much lower starting position, the United States achieved a much faster growth rate and hence overtook the United Kingdom's GDP per head by 1913. Similarly, Germany was growing at a faster rate than the United Kingdom after 1870.

Growth rates in Europe shown for the twentieth century have been much affected by the war years. Thus, the United States was able to grow consistently faster than the United Kingdom and Germany up to 1950. This phenomenon of being overtaken by faster growing competitors was to be again seen when compared to European nations in what was to be described as the golden period of economic growth from 1950 to 1980. Thus, in the period 1950 to 1979 Germany's GDP per person went from 61% of the United Kingdom to 116% [11].

The surprisingly low growth figure shown for Germany since 1989 is a result of the reunification of East and West Germany in 1990 with the lower GDP per head of the East lowering the overall GDP per head figures. The results for 1950 and 1989 are based on West Germany alone.

The key questions are: Why has the United Kingdom shown slower growth? Can this performance be reversed? Crafts' analysis highlights the importance of growth in what is described by him and other economists as total factor productivity (TFP). This is a measure of the efficiency of an economy in its use of labor, capital, and land. It is shown that other economies, such as the United States, were able to increase their TFP over a long period and this enabled them to outperform the U.K. economy from 1870 onwards. This improvement in TFP is largely attributed to the more successful adoption of new technology and the changes that went with it (such as mass production). New technologies were initially electricity, chemicals, and steel but later involved information and communication technology (ICT). Again, several factors are highlighted that potentially effected this relative performance; these include:

- Britain's heavy investment in steam-power based technology, making it less willing to adopt other motive power (see Figure 1.8). Up until 1950, the established industries of coal, cotton, iron, and shipbuilding dominated Britain's economic thinking.
- The weakness of Britain's education system in producing technically qualified managers and staff.
- A craft-based employment structure that was likely to be resistant to change to mass-production techniques (Fordism).
- Britain's long-term commitment to free trade that suited its established businesses but weakened the growth of new industries if other countries led the development (examples are automobile manufacturing and chemicals).

These issues are deeply embedded in the United Kingdom's culture and have been analyzed by several observers. Freeman's work on innovation pointed to the United Kingdom being the birthplace of the Industrial Revolution as

Figure 1.8 A dirty business: Britain's coal-powered steam railways lasted until the 1960s. (Source: Kevin Lane Collection.)

not a simple matter of happenstance but due to the existence of a wide range of factors favoring innovation [12]. Elbaum and Lazonick highlighted the institutions that operated in Britain under a strong belief in a free-market economy where the "invisible hand" of the market guides development and therefore leads to the resulting factors shown [13]. This belief in a true free market has been a significant factor within the United Kingdom since the start of the industrial revolution although the distance of U.K. colonies from central review enabled monopolistic actions in the exploitation of their resources. The belief in free markets contrasted with the many competing economies where the existence of large (potentially monopolistic) organizations enabled a growth in productivity beyond what was achieved in the United Kingdom [14].

While the importance of these and other factors have been debated for some time, the issues have not gone away. Many observers believe that the United Kingdom still suffers from a weakness in technological resources and management [15]. This issue was widely explored throughout the period of this study. Some authors claimed that Britain's entire culture was geared towards an Oxbridge educated elite who were well suited to running an empire but notably weak at dealing with technology issues. This tension is explored by C. P. Snow in his "two cultures" debate [16].

1.3.3 U.K. Politics: 1950 to 2000

In this period, the United Kingdom's politics was dominated by the two main political parties, Conservative and Labour. They broadly took turns of around 10 years at running the country. The Conservatives with their professed belief in the free market were in power from 1951–1964, 1970–1974, 1979–1997, and beyond 2008. Labour, with their support for state ownership and control, were in power 1950–1951, 1964–1970, 1976–1979, and 1997–2008 (although post 1997, the Labour party largely abandoned its support for state ownership of industry).

It might be expected that the two parties with their profoundly different political ideologies would produce dramatically different approaches to U.K. industry. While they often seemed at odds over political philosophy, between 1950 and 1980 they adopted a reasonably consistent policy of supporting the state monopolies in areas such as utilities and telecoms while keeping a tight control of their funding which many observers believed restricted their growth and development. The reasons for this are debatable but the emergence of a consistent policy (at the time referred to as "Butskellism" (a combination of the names of the Conservative minister Rab Butler and the Labour leader Hugh Gaitskell) was possibly a result of a shortage of finance (due to the high borrowings to finance World War II) and the predominance of other issues relating to

Britain's retreat from colonialism and subsequent entry in 1973 into what was to become the European Union [17].

The Labour party believed in significant state investment but was heavily committed to a wide range of social programs (such as building the National Health Service (NHS)) as well as the considerable costs of supporting recently nationalized industries such as coal and the railways. The Conservative party was nervous of massive state investment as it believed this would distort the market economy and be wasteful. Throughout the period from 1950 to 1970, the United Kingdom struggled to invest adequately in infrastructure and what investments were made were in established areas such as electricity generation. Few attempts were made to harness the considerable power of the City of London to provide capital for Britain's infrastructure. The City of London continued to operate in its traditional detached manner with investments going into property, overseas investments and private companies.

To many observers, the most significant political event in the period was the election of the Conservative government in 1979 under the leadership of Margaret Thatcher. This government carried through a series of radical, neoliberal reforms with an aim to reduce the traditional power of the trade unions and to move many enterprises to private ownership where (in the opinion of the government) they could flourish. Receipts from the sale of state enterprises provided a chance to avoid major tax rises while increasing some public spending. The privatization of many state-run enterprises included the national gas, electricity, and water industries plus the national airline. The telecoms network operator (BT) was a key part of this privatization process (which left postal services untouched in public ownership). Many other assets were also removed from state ownership including 2 million publicly owned dwellings. The assets were sold off to the British public at often attractive prices with the aim of creating a large property-owning class while the receipts were used to reduce taxation (for a while). The benefits (or otherwise) of "Thatcherism" continue to be debated [18] (see Figure 1.9), but their effect was influential throughout the main economies of the world with, for instance, most European countries privatizing their telecoms networks in the wake of the United Kingdom's example. Even when a Labour government was elected in the United Kingdom in 1997, it made little attempt to reverse the main policies of the Thatcher government.

1.4   The Economics of Telecoms

It is now time to look at economic theory and how it applies to the telecoms industry. I will assume that the reader is (like me) not a trained economist and I will only cover economic theory to the extent that it is relevant to this work.

SI MONUMENTUM REQUIRIS, CIRCUMSPICE.

Figure 1.9 Not all observers were impressed by Margaret Thatcher's reforms. (Source: Nicholas Garland.)

1.4.1   Waves of Development

How technology affects society is a key area for economic research. Seminal works in this area have been published in the last hundred years. One of the first to observe long-term economic cycles was the Soviet economist Kondratieff [19], who in 1926 identified the long time cycles (of around 50 years) that apply to economic indicators such as bond yields and commodity prices. Many economists have subsequently explored this phenomenon and suggested possible reasons for these waves [21]. Of particular relevance was the analysis of Schumpeter, who argued that a significant factor in these cycles was innovation. Thus, the economic development of countries could be characterized as driven by a series of waves, which were, in turn, driven by technical innovations. Proponents of this view often list 5 waves [22] (with approximate starting dates):

1. 1771: The industrial revolution. The age of waterpower, canal transport, and textile manufacturing.
2. 1829: The introduction of steam power. The age of railways and ironworks.
3. 1875: The age of steel and heavy engineering.
4. 1908: The age of electricity, oil, automobiles, and mass production.
5. 1971: The development of ICT.

My research found this to be a factor of considerable relevance when considering the expansion of a country's telecoms network. Schumpeter [21] postulated the concept that the waves of major technology development by a process of creative destruction would enable the new technology to improve the performance of society. An example of this is the introduction of coal-powered steam engines that drove a major phase of the industrial revolution.

In the postwar era, there were major technological developments that, although perhaps not on par with the invention of steam power, were nevertheless very significant in their effect on the telecoms industry and beyond. I would include in this list:

- The invention of the transistor;
- The development of the planar epitaxial process, leading to the introduction of the silicon integrated circuit;
- The invention of fiber-optic data transmission;
- The development of mobile phones;
- The development of the internet and the World Wide Web.

Here I refer to these changes as Schumpeterian "miniwaves." All of these had a major effect on the telecoms industry both by spectacularly increasing the demand for infrastructure and by drastically changing how this infrastructure could be provided (and its consequent cost).

The creative destruction described by Schumpeter does not necessarily just apply to national economies. Other economists have applied the same concept to large enterprises arguing that, if not challenged, can lead to economic inefficiency [23]. Some of this thinking was later to be applied to the privatization of the national telephone operators adopted by the United Kingdom and elsewhere. However, the analysis also showed that privatization per se did not achieve creative destruction. In many cases, the breaking up of these large enterprises is seen as the necessary action to facilitate technological innovation.

There has been considerable research into how industries and firms respond to innovation. Freeman and Perez identified categories of innovative change [24]. They sought to differentiate between inventions (such as most of the items listed above) and innovations where the inventions are developed and applied to change the total situation in which the industry operates. Their argument is that it is not the invention itself that is the most significant factor but the actions that surround it to turn it from an idea into something that can seriously affect society. At this point, it could be considered an innovation. The technological developments listed above evolved into a series of innovations that facilitated the automation and then digitalization of the telecoms network. These changes described in this work add up to a "change in techno-economic paradigm" (as an economist would term it) where both the demand for telecoms and the costs of supplying it went through a major transition with consequences (such as the growth of social media), which were almost impossible to predict.

Others, notably Malerba [25], looked at how firms respond to such change. However, little of this research has been undertaken looking into a situation of a monopoly customer, which is typically the position in the 1950s that the national, government-controlled posts, telephone, and telegraph (PTT) authority enjoyed in most developed countries. The issue arises from the existence of a monopoly customer who insists on keeping a significant control over the entire telecoms network and its design, configuration, and deployment. This means that there is low appropriability of innovation (as Malerba would describe it) with a possible response of collaboration. By this, he implies that innovations would be unlikely to become a significant factor in generating success for a supplier to the telecoms network of a country, unless they were of enormous significance (such as the invention of semiconductors). However, in most countries, notably throughout Europe, the supply industry for telephone networks was considerably smaller and weaker than the network operators they supplied (and hence unlikely to build a strong commercial position by innovation). A notably different situation existed in the United States where the AT&T company was both a monopoly supplier and operator of telecoms.

The example of the invention of the transistor by AT&T in its Bell Labs research center in 1947 shows how a major innovation could arise within the telecoms industry. This was the key innovation in the development of electronics affecting this period (see Figure 1.10). However, due to antitrust concerns and a general wish to provide a broad social benefit, this invention was made widely available, hence reducing the appropriability of this innovation. In more basic terms, it is hard to believe that Bell ever made the sort of financial gain that you would expect from one of the most significant innovations of the twentieth century. However, the financial and social benefits that this invention spread throughout the world probably far exceed anything that could have happened if the invention was strictly controlled by AT&T [26].

Thus, it can be argued that a firm whose customer for most of its output is the national PTT would not use innovation to achieve market advantage but rather as a necessary overhead to be managed as a cost of doing business. Under these circumstances, innovation was mainly driven by the internal dynamics of the PTT. This was largely motivated by direct and indirect political pressure, which was passed on to the telecoms network operation. These state-controlled monopoly operators were thus large, bureaucratic, and often lethargic. Later in the period, innovations (such as the development of the internet) were largely driven by organizations outside the traditional PTTs and their suppliers.

1.4.2 The Effect of Globalization

A major factor that was to affect the telecoms industry as it developed in the postwar period could be simply summarized as globalization. As will be seen,

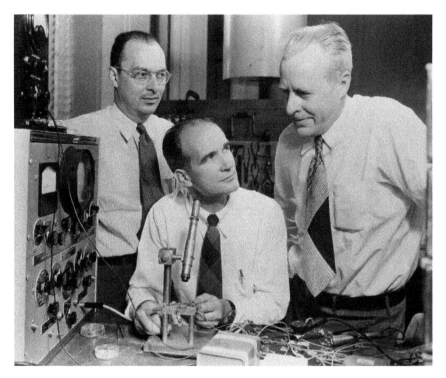

Figure 1.10 William Shockley, John Bardeen, and Walter Brattain, inventors of the transistor at Bell Labs. Nobel Prize winners of 1956. (Source: AT&T.)

the telecoms industry of the prewar period was, to a considerable extent, insular. Each major economy would develop its own network, mainly using local suppliers. This was set to change dramatically from the 1970s as the trends in globalization began to affect the industry. The phenomenon of globalization was not new and had been observed from the beginnings of the twentieth century as it affected major industries such as cotton manufacturing. The basic concept looks at the stages by which a product innovation is introduced into a global market. Several U.S.-centered researchers particularly looked at how U.S. companies had increasingly moved production out of the United States as their innovations matured [27]. We can identify 4 stages:

1. A product innovation is introduced by a company operating in a major market where it has both the knowledge of the market and the technical and financial resources to innovate.

2. If the innovation is successful, the company introducing the product scales up production and exports its innovation to other markets.

3. Competitors begin to emerge and erode the financial advantage of the original innovator.

4. The pressure on costs leads to production being moved to locations where lower-cost manufacturing can be achieved.

The relatively stable political and economic situation after 1945 led to a rapid increase in the globalization process. Initially, simple mass-produced items (such as the transistor radio and telephone handsets) moved to a predominately Far East manufacturer, but by the end of 2000 much more sophisticated products were subject to the same trends. This phenomenon formed the backdrop for considerable political debate in the postwar era [28], which persists to this day (see Figure 1.11). The key issue revolves around whether it is appropriate to resist this phenomenon by protectionism and subsidy of local manufacturing to protect local employment or to embrace it as a way to increase the overall wealth of a country [5].

As the fifth wave of innovation, ICT took shape in the 1970s, and observers began to realize that this could have a big effect on the dynamics of global growth. Historically, it was implicitly assumed that only highly developed countries would benefit from major innovations. However, the nature of ICT with its relative low capital cost and highly mobile know-how (often provided

Figure 1.11   A contrarian view of the benefits of globalization. (Source: Tony Biddle.)

by a foreign-educated workforce) posed the possibility of developing countries leapfrogging the classic development process [28]. Thus, by 2000, some developing countries were building mobile phone networks without having first established a comprehensive landline telephony system [29]. This phenomenon would increase the challenge to the role of a traditional supplier to a national telecoms network.

1.4.3   Increasing Personal Wealth

What should not be overlooked is the effect of growing wealth throughout the world in the 50 years of this study. This created a latent market for telecoms in all its manifestations for both individual consumers and organizations together with the funds required to develop the infrastructure. Cultural developments in this period, such as digital media and the increasing mobility of people, conspired to drive demand for telecoms dramatically upwards. So, while the world population more than doubled from 1950 to 2000 (to 6 billion), the number of fixed-line phones increased by more than a factor of 10 (to 16 per 100 people) and the number of mobile phones rose from nothing to 12 per 100 people.

As can be seen from the graph in Figure 1.12, the number of phones in use followed much the same path in most developed countries with fixed lines increasing steadily until 2000. By then, the dramatic growth in mobile phones that started in 1990 was well under way and, by 2010, there were more mobile phones in use than fixed lines. In some countries (such as the United Kingdom), there was even more than one mobile phone per person.

From 1950 onwards, the growth in telephony, particularly between countries, would demand an increase in standardization and regulation that would be facilitated by the strengthening of international organizations. The era of

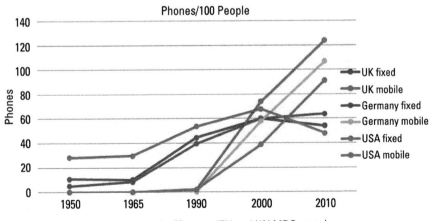

Figure 1.12   Phones per 100 people. (Source: ITU and UN MDG stats.)

decolonialization would lead to the liberalization of trade in telecoms equipment, which, in turn, would disrupt existing supply arrangements. The growth in usage of telephony would also lead people to begin to question the monopolistic behavior of telecoms network operators.

The political climate of the 1980s, which favored privatization and demerger, coincided with major market opportunities relating to the start of spectacular growth in mobile telephony and data communications. These major innovations were facilitated by the political climate and the creation of international standards that enabled interoperability and enabled new organizations to enter the market to accelerate the rapid acceptance of these new innovations.

The period covered was one of major mergers and acquisitions and the telecoms supply industry went through its fair share of changes. As will be shown, the period saw the U.K. telecoms industry go from being a major exporter to a large net importer. By 2006, there was no U.K.-owned top tier manufacturer, with the United Kingdom largely dependent on imported products and technology.

1.5   The World's Telecoms Industry

While the detailed role and structure of the telecoms industry are discussed in later sections, it is important at this stage to understand how the inherent nature of telecommunications drives how the industry operates. In common with most infrastructure, the industry engages in large capital investments that must operate with a high reliability over a life cycle that usually exceeds 20 years.

For the worldwide telecommunications industry, one significant factor that applied to most countries at the start of this period is the existence of one monopoly supplier of post and telecoms services in each country (Figure 1.13). In the 1950s, this was usually the PTT. This was true of most countries in Western Europe [30] as well as Japan where the national telecoms operator had been run by AT&T in the period after World War II but had been handed back to Japanese state control in 1952.

In major economies in the 1950s, the supply of telecoms equipment was invariably carried out by local suppliers. In most countries, these were private companies with often one or more dominant suppliers who had an extremely close relationship with the national PTT. However, the United States was an exception to this. As a huge private company, AT&T enjoyed a virtual monopoly of telephony in most areas and used its in-house manufacturing arm (Western Electric) to meet most of its product needs. However, although AT&T was always a private company, the strategic importance of telecommunications meant that the U.S. government maintained an extremely close watch over the company. Post was handled by the U.S. Post Office, a separate government concern

Figure 1.13   The PTT, often a matter of national pride, Post and Telegraph Offices, Flinders Street, Townsville, Australia, circa 1890. (Source: Queensland State Archives.)

and telegraphy largely by Western Union, another independent company (which had between 1909 and 1913 been controlled by AT&T). Throughout this period, all countries had to address the same issues of technological change and the ownership and control of its PTT that arose in the United Kingdom. How their actions compared with the United Kingdom and the resultant effect on their supply industry are thus key parts to this analysis.

In the United Kingdom, the PTT was the General Post Office (GPO) and its telecoms activities were privatized as BT in 1984. The GPO went through significant changes as it evolved from being an office of the U.K. Crown (a department of government) to being split into separate business entities with its telephone network, BT, becoming quoted on the U.K. stock exchange. However, privatization was not of itself likely to reduce the monopolistic power of the telecoms network and raised in the United Kingdom and elsewhere issues of regulation and the need to inject competition into the marketplace. The privatization of the U.K. telecoms network was a key event in the political history of the United Kingdom and influential throughout the world in how governments related to public utilities and other monopolies [18]. As will be shown, this privatization took place at a time when the U.K. telephone network was undergoing major technological change. While a lot of discussion was undertaken about the effect of the privatization of BT on consumers, there has been relatively little discussion about the long-term effect on its telecoms supply industry. The U.K. telecoms supply industry was affected by the twin challenges of technological change and the privatization of its monopoly customer. These changes significantly affected the industry's ability to support the U.K. telecoms network and win export business.

In the case of the United States, the breakup of the Bell Telephone monopoly was a major event that also happened in 1984. The break was a result of antitrust concerns that had been a running issue between the U.S. federal government and AT&T since it established its dominant position in U.S. communications in the early 1900s [31]. The effect of this action has been extensively explored elsewhere and provides an interesting comparison to this analysis of the U.K. telecoms industry [32]. It will be shown that the 2 countries whose telecoms manufacturing sectors performed worst in this period (measured in terms of relative import and export performance in Appendix A) were the United Kingdom and the United States.

The introduction of mobile phones, cable TV, and the subsequent development of the internet from the 1980s did dramatically broaden the competitive horizon of the telecoms industry but to a considerable extent the telecoms supply industry in the United Kingdom did not benefit from these developments as a major new market opportunity. This response is consistent with the findings of Christensen [33] in his review of the Computer Disc Storage Industry whereby incumbent suppliers find it hard to recognize and address market developments that might ultimately threaten their core market. However, as will be explored later, these developments were much more significant and represented major changes as envisaged by Schumpeter and others in the process of creative destruction that drives economic development [22].

Given the role of the PTT in specifying and deploying its telecoms network, trade in telecoms equipment prior to 1950 was often seen in a political context. In particular, the United Kingdom would, as a matter of course, deploy U.K. telecoms equipment throughout its colonies and countries of influence. Similarly, U.S. telecoms practices would be extensively followed in Latin America.

How the U.K. telecoms industry developed was affected by the major technological changes that occurred and the political actions that went with it. Having already reviewed the economic framework for these changes, we can now start to see how sectors of the industry reacted. To help the analysis, I have looked separately at three major subdivisions of the telecoms sector: voice switching/transmission, mobile telephony, and data communications. While these sectors interrelate, they can be considered separately.

Having plotted the development of these subsectors I will then look at how the U.K. telecoms supply industry reacted to these changes compared to other countries. The results will form the basis of my conclusions, which give an overview of the resultant outcome and draw out what messages can be gained for the wider U.K. economy.

1.6   The Effect of Technology

If we define telecoms as the transmission of information over distance, its foundation predates the discovery of electricity, let alone the development of electronics. Early examples would be the generation of smoke signals and sending high-intensity sounds across a distance. Pioneering work by Claude Chappe in northern France led to the development of an optical transmission system using shutters. In 1791, this was successfully demonstrated and given the name of "telegraphe" [34].

One of the earliest organized transmission arrangements was set up by the U.K. admiralty to transmit data from its main base in Portsmouth to its headquarters in London at the time of the Napoleonic Wars in the early 1800s (Figure 1.14). An ingenious series of telegraph stations could relay complex messages over the 60-mile distance in minutes. It can thus be argued that data

Figure 1.14   Chadley Heath Semaphore Tower, Surrey, United Kingdom. (Source: Nigel Richardson CC by-SA 2.0.)

transmission (such as the above example) preceded speech transmission, which required electrical connections to operate. While the two forms of telecoms interplay, it was mainly the development of telephony that drove the industry until the development of computers and the internet brought data communications back to the fore. For this reason, I shall largely consider the topics separately and initially concentrate on speech transmission (Chapter 2).

Since the first demonstration of a telephone call made by Alexander Graham Bell in 1876 [35], there was little doubt that there was a market for the telephone network and all that went with it (Figure 1.15). What was needed were the many incremental technical developments to make this feasible and the capital to roll out the network so that eventually everywhere would have access to telephones. However, most observers at the time thought the telephone would only be used by rich people and large organizations.

The basic concept of the telephone call did not change much between the original demonstration and now. However, two fundamental issues needed to be resolved: transmission and switching.

Clearly, as the distance between the sender of a message and its receiver increased, then the quality of the call would fall off without some form of amplification. Also, unless one envisaged a telephone network where everyone was

Figure 1.15 Alexander Graham Bell, inaugurating the 1,520-km telephone link between New York City and Chicago on October 18, 1892. (Source: http://www.americaslibrary.gov/jb/recon/ jb_recon_telephone_1_e.html. Public Domain.)

connected to everyone else with an enormous number of wires, then some form of switching system would be required. Initially, in the case of switching, a simple manually connected plug board type telephone exchange was adopted and proved to be quite effective with manual telephone operators in use right up to the 1960s.

Transmission was a bigger issue but the invention of the thermionic valve by Fleming in 1904 enabled amplification of speech and hence long-distance calls. It also made possible the multiplexing of signals so that multiple speech channels could use a single pair of copper wires. (The cost of copper was a major issue with early network development.) The value of multiplexing is easy to overlook but was crucial to the development of communications networks. Its use was initially pioneered in the development of telegraph networks in the nineteenth century using mechanical devices that sent signals from several lines in sequence over a single link (see Section 2.1) to achieve a simple form of multiplexing (called time division multiplexing (TDM)). Such a technique did not easily facilitate the transmission of multiple speech signals. Thermionic valves enabled a more sophisticated method of multiplexing whereby each signal is added to a different frequency carrier signal on the single line (frequency division multiplexing (FDM)). Thus, a reasonably effective trunk network was rolled out in most countries by the beginning of the twentieth century.

Between 1950 and 2000, a number of what I earlier called Schumpeterian miniwaves hit the industry as electronics was developed.

1.6.1   The Transistor

As already mentioned, the first and perhaps greatest miniwave was the invention of the transistor (Figure 1.16). This took place in Bell Labs in America, a subsidiary of Western Electric, the supply arm of AT&T (the then-monopoly supplier of telephone systems in the United States) in 1947 [36]. It took until the 1950s for this invention to start to affect the telecoms industry. William Shockley, John Bardeen, and Walter Brattain, the named inventors, were awarded the Nobel Prize in 1956.

Bell Labs had ample funding to support long-term development into advanced amplification devices and saw an obvious market need for such technology. Thermionic valves were large, consumed a lot of energy, and required a high degree of maintenance as the mean time between failure was only a few thousand hours. All these factors could be greatly improved even by the early discrete transistors that were available in the 1950s and so these were initially put to use in transmission systems on the network. Thus, from the 1960s, the existing copper transmission lines used for telecoms achieved substantial increases in signal-carrying capacity using multiplexing operating at increasing frequency (which enabled more signals per line).

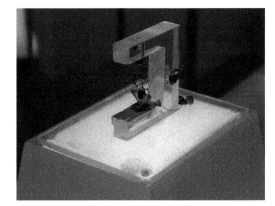

Figure 1.16 An early transistor made in Bell Labs. (Source: Windell Oskay, oskay CC-2.0.)

Electronic switching was a much bigger issue. The initial manually operated switchboards using operators were obviously labor-intensive and difficult to scale. As early as 1899, an electromechanical alternative had been developed. One of the first organizations to do this was a Kansas City undertaker who, getting fed up with operators overhearing calls and routing business to competitors, came up with a mechanical way of removing operators and hence keeping his business from being stolen. Thus, the Strowger system was developed [37] that consisted of a series of electromechanical rotary switches, which would switch lines based on electrical pulses from the mechanical dial fitted to the telephone. This pulse dial arrangement lasted again well into the 1960s in the United Kingdom. Electromechanical telephone exchanges were rolled out throughout the world and proved to be superior in terms of labor productivity than employing telephone operators. However, they were mechanically intricate and required a lot of maintenance. The United Kingdom adopted the Strowger system because it was proposed by a number of the indigenous suppliers in the 1920s who had license arrangements with U.S. manufacturers [38]. Superior alternative systems had been developed and were widely adopted, particularly in Europe. Later the development of the Crossbar switch used electromechanical switches arrayed in a matrix that facilitated a much more efficient switching of signals. The GPO in the United Kingdom stuck to its original choice of the Strowger switch (Figure 1.17) and only reluctantly started buying Crossbar exchanges in the 1960s. By then, it was clear that they were some way from being able to install fully electronic exchanges and the deficiencies of Strowger exchanges were becoming painfully obvious.

It was widely expected that with the invention of the transistor some form of solid-state electronic switching could be used instead of the electromechanical arrangements used throughout the world in the 1950s. However, this proved far harder to crack than was originally envisaged. In the 1950s, the GPO

Figure 1.17   Part of a Strowger exchange. (Source: BT Archives.)

had been optimistic about rapidly replacing its Strowger network with fully electronic switching, but many country PTTs took a more cautious approach.

While the switching itself, using semiconductors, proved much more difficult than had been envisaged, a considerable amount of progress was made in the 1950s with the overall management of the switching network, which was almost as big an issue. Alongside the development of Crossbar exchanges suppliers had begun to develop electronic call-management systems, for what is often termed the Register of an exchange. Initially, hardwired electronics were used, but later this evolved to software control (often referred to as stored program control (SPC)). Along with this was also early work carried through with the use of a hybrid switching system, which would use electronic control but reed relays for the actual switching. This could achieve some of the advantages of electronic switching system without the need to solve the switching problems of semiconductors that were still to be overcome.

1.6.2   The Silicon Integrated Circuit

Introducing a full electronic switching system in a telephone network required a second Schumpeterian miniwave, the introduction of the silicon integrated circuit. The invention of the transistor had led to the creation of numerous businesses that licensed the transistor patents, initially in the United States but also in Europe and Japan. Two such licensees were Texas Instruments and Fairchild. These companies helped to create the next wave with the development of the silicon integrated circuit. Jack Kilby working at Texas Instruments was able to demonstrate a simple integrated circuit in 1958. This used germanium as the semiconductor material as did most transistors at that time. However, soon after Robert Noyce independently demonstrated a similar development at Fairchild but this time using silicon as a base material. This showed the route by which large-scale digital signal processing could be instituted in an application such as a telephone exchange. The transistors developed from the original invention at Bell Labs were originally discrete devices that performed a basic amplifying process analogous to that of the earlier triode thermionic valve. The early developments used germanium as the base (substrate) material with a junction individually placed on top of a small piece of germanium to achieve the transistor function. However, intensive research by several device manufacturers, notably Fairchild and Texas Instruments of the United States between 1958 and 1960, developed the planar epitaxial process using silicon as a base substrate. This process enabled devices to be laid out on the surface of the substrate material with complex patterns created by a photographic process. Hence, it could be seen that multiple devices could be created on the same substrate. By applying metallization layers, the interconnection of these devices could be integrated onto the base substrate of the device (hence, integrated circuit) (Figure 1.18). While the original integrated circuits only included a few devices, it was clear that the technology could expand to cover an increasing number of devices. (By 1990, this number was over 1 million.) The silicon integrated circuit benefited from the application of Moore's law [39]. This remarkable prophecy, which held for over 40 years, forecast that the density of gates on an integrated circuit would double every 18 months (later increased to 24 months). This would lead to a dramatic reduction in the cost of semiconductors every year (Figure 1.19).

Figure 1.18   Early integrated circuit. (Source: Angeloleithold CC-SA 3.0.)

Figure 1.19   Intel 486 processor, launched in 1989 with more than 1 million transistors. (Source: byzantiumbooks CC 2.0.)

Semiconductor switching could thus be achieved by digitizing the incoming analog signals from voice calls, passing them through a digital switching network, and then converting the signal back to an analog voice signal at the receiver's line. This approach also offered a substantial improvement in reliability. Initially, the costs of equipment to process calls digitally were much higher than existing electromechanical exchanges. However, it was only a matter of time before Moore's law changed this calculation. The potential benefits of a full digital telephone exchange were widely appreciated and many telecoms companies set about developing electronic exchanges in the 1970s.

Behind this top-level development would also go the requirement for a substantial reengineering of the whole telephone network for digital transmission. This was eventually confirmed by the adoption of Integrated Services Digital Network (ISDN) as set out by the Consultative Committee for International Telephony and Telegraphy (CCITT), the international trade body, in 1988. This was implemented by the GPO progressively from the 1970s. As with many innovations, they are much easier to implement if you are starting from scratch than to carry them through into an operating system. Applying these standards to a telephone network of considerable complexity required a great deal of ingenuity and management.

1.6.3   Fiber Optic Cables

Another Schumpeterian miniwave that emerged in the 1980s was the introduction of fiber optic transmission. Although the signal-carrying capability of the basic copper coaxial wires used in transmission had been greatly enhanced by multiplexing techniques, there was a likely ceiling to the improvement possible driven by the maximum signal frequency (bandwidth) of the cables. Thus, there had been a continuing search within the industry to increase the bandwidth of transmissions, which would enable more separate signals to be carried. Fiber optic transmission offered a route to reduce the costs of transmission within the network, since using light as a carrier instead of an electrical signal dramatically increased the bandwidth available for transmission. There were numerous candidates to increase bandwidth that were actively explored. These included:

1. Higher-frequency transmission wires, which still used conventional electronic signals;

2. High-frequency microwave radio transmission, using line-of-sight transmission towers;

3. Waveguides, which used high-frequency microwave radio signals down transmission tubes;

4. Satellite transmission, using microwaves beamed to and from a satellite;

5. Fiber-optic transmission.

All these approaches were developed and trialed, and many are still in use, but the use of fiber-optic transmission cables proved the most appropriate for many applications, including undersea transmission. They provided a spectacular improvement in the amount of information that could be transmitted down a cable where, instead of using electrical signals in a copper cable, modulated light is sent down a glass fiber (over a thousand times greater data transmission could be achieved compared with copper cables).

The idea of using optic transmission of voice signals was demonstrated by Alexander Bell in 1880, but the real breakthrough was made in 1966 at the United Kingdom's STL lab where it was shown that dramatic improvement in transmission could be achieved by using very high purity glass (Figure 1.20). Corning Glass Works of the United States produced an optical fiber transmission cable in 1970. The parallel development of the semiconductor laser diode enabled light to be modulated at a high bandwidth. This led to the first commercial fiber-optic transmission system in 1975 with a bit rate of 45 Mbps and a repeater separation of 10 km.

As the cables were of a similar size and physical characteristics to copper cables, many of the well-proven techniques for laying and maintaining cables

Figure 1.20   Fiber that is the same size as copper cables but has thousands of times the capacity. (Source: Asharkyu, Shutterstock ID 1358176214.)

could be used for the new fiber cables. This was a much simpler way to expand the capacity of the telephone network compared with alternatives such as microwaves or satellite transmission. This was particularly important in transmission across the seas, which for reasons dating back to the British Empire, the United Kingdom maintained a strong interest. The adoption of fiber-optic communications for undersea cables culminated in the installation of TAT-8 between the United States, Britain, and France in 1988. This effectively replaced satellite communications for transmission of voice, data, and video signals between the continents [40]. By 2000, fiber optic cables had progressed from an original data transmission rate of 45 Mbps to over a terabit per second (1,000,000 Mbps).

The dramatic increase in transmission capacity, first using copper coaxial cables and later fiber optic cables, became a major stimulus for the introduction of digital services such a fax machines (introduced in the late 1970s) and early data communications (1980s onwards). Initially transmission of data simply used the voice channels with a modem to transmit the data onto the channel. The speed of such modems progressively increased from a few hundreds of bits per second to over 9,600 bps. This speeding up of data transmission led to the demise of the old established (and separate) telegraph and telex networks with their much slower transmission rates, as described later.

As data traffic became increasingly important, it became possible to look at the transmission in a different way whereby individual pieces of data would be sent down the network each carrying its own address. This is usually termed packet switching. The funded development of this approach by the U.S. government Advanced Research Project Agency (ARPA) led to the creation of what became known as the internet. Thus, we are now at a point where the digital switching of voice signals is only part of the telecoms network which would usually be regarded under the heading packet switching networks. This topic is explored in more detail in Chapter 4.

The development of semiconductors and fiber optic transmission made it possible for the market for mobile phones to develop. As will be discussed in Chapter 3, while the basic concepts were explored earlier in the twentieth century, it was only in the 1980s when the cost of data processing and transmission had been dramatically reduced that the mass consumer market that we now know was able to develop.

1.6.4 The Effect on the Telecoms Supply Industry

The introduction of digital switching and transmission was a major disruption for the supply industry since their factories at the time were virtually exclusively geared to manufacturing small mechanical parts and assembling them into electromechanical switching systems. As such, these facilities were highly labor-intensive and required a substantial infrastructure relating to mechanical engineering. These factories also had a secondary potential use in that, in times of war, such factories could be repurposed to produce munitions, military communications, and similar items for warfare [41]. It is widely felt that some of the delays of the U.K. telecoms industry to wholeheartedly adopt digital electronics was because of the reluctance to stop these factories from operating. In the time-honored fashion, politicians and industrialists wanted to be ready to fight the last world war.

With the expected widespread adoption of electronic switching and the increasing use of electronics in data transmission, the electronic supply industry had to face the issue that they had many electromechanical factories that were obsolete. In practice, it was easier to shut down these factories and build up capabilities in electronics in other locations. This is explored in later chapters.

From an employment/governmental point of view, the industry was subject to a dramatic decline since all the electromechanical production would be rapidly terminated and the corresponding growth in electronics would, to a fair extent, come from externally supplied components, many of which were imported. This would have been the case even if the industry could develop successfully its own digital switching capability.

The development of the mobile phone is perhaps the most significant development in telecoms in the second half of the twentieth century. As the capability of semiconductors increased and their costs reduced, the use of radio transmission rather than wired connection was an obvious possibility, which was explored widely from the 1950s onwards. As is shown in the following chapters the existing U.K. telecoms supply industry did not very actively pursue this obvious opportunity. The capability of the improving productivity and cost of digital electronics made the mobile phone an attractive possibility as an alternative to fixed-line telecoms, but there is little sign that this was identified as any form of business opportunity by the main suppliers. In line with the work of Christensen, this major change in market was mainly addressed by newly created businesses except for the spin out of Cellnet (later renamed O2) from BT.

References

[1]   Owen, G., *From Empire to Europe*, London: Harper Collins, 1999.

[2]   Hutton, W., *The State We're In*, London: Vintage, 1996.

[3]   Hamilton-Paterson, J., *What We Have Lost: The Dismantling of Great Britain*, London: Head of Zeus, 2018, p. 203.

[4]   Zysman, C., et al., "Why Manufacturing Matters: The Myth of the Post-Industrial Economy," *California Management Review*, Vol. XXIX, No. 3, 1987.

[5]   Tyson, L. D., *Who's Bashing Whom?*, Institute for International Economics: Washington, D.C., 1992.

[6]   Alam, M., et al., *Analysis of the Dot-Com Bubble of the 1990s*, Manhattan, KS: Kansas State University, 2008.

[7]   Landes, D. S., *The Unbound Prometheus: Technology Change and Industrial Development in Western Europe from 1750 to the Present*, Cambridge, U.K.: Cambridge University Press, 2003.

[8]   Crafts, N., *Forging Ahead, Falling Behinds and Fighting Back, British Economic Growth from the Industrial Revolution to the Financial Crisis*, Cambridge, U.K.: Cambridge University Press, 2018.

[9]   Broadberry, S., *The Productivity Race: British Manufacturing in an International Perspective 1850-1990*, Cambridge, U.K.: Cambridge University Press, 1997.

[10] Lawson, K., "On the Nature of Industrial Decline in the UK," *Cambridge Journal of Economics*, 1980, pp. 85–102.

[11] Crafts, N., *Forging Ahead, Falling Behinds and Fighting Back, British Economic Growth from the Industrial Revolution to the Financial Crisis*, Cambridge, U.K.: Cambridge University Press, 2018, p. 2.

[12] Freeman, L., et al., *As Time Goes By: From the Industrial Revolutions to the Information Revolution*, Oxford, U.K.: Oxford University Press, 2002.

[13] Elbaum, B., and W. Lazonick, "The Decline of the British Economy: An Institutional Perspective," *The Journal of Economic History*, Vol. 44, No. 2, 1984, pp. 567–583.

[14] Kirby, M. W., "Institutional Rigidities and Economic Decline: Reflections on the British Experience," *Economic History Review*, 1992, pp. 637–660.

[15] Hauser, H., *Current and Future Role of Innovation Centres in the UK*, London: Department of Business, Industry and Skills, 2010.

[16] Snow, C. P., "The Two Cultures," *The Rede Lecture,* Cambridge, U.K.: Cambridge University Press, 1959.

[17] Studler, D. T., "Orders and Eras in Post-War Britain," *The Forum,* Vol. 3, No. 3, 2007.

[18] Edwards, C., "Margaret Thatcher's Privatization Legacy," *Cato Institute,* Vol. 37, 2017, pp. 89–101.

[19] Kondratieff, W. F., "The Long Waves in Economic Life," *The Review of Economics and Statistics,* 1935, pp. 105–115.

[20] Solomou, S., *Phases of Economic Growth, 1850–1973: Kondratieff Waves and Kuznets Swings,* Cambridge, U.K.: Cambridge University Press, 1990.

[21] Schumpeter, J., *Business Cycles,* New York: McGraw-Hill, 1939.

[22] Freeman, S., et al., *Economics of Industrial Innovation,* Oxford, U.K.: Routledge, 1997.

[23] Fogel, K. M. R., "Big Business Stability and Economic Growth: Is What's Good for General Motors Good for America?" *Journal of Financial Economics,* 2008, pp. 83–108.

[24] Perez, F., et al., "Structural Crises of Adjustment, Business Cycles and Investment Behaviour," in Dosi, G., et al., *Technical Change and Economic Theory,* London: Pinter, 1988, pp. 38–66.

[25] Marleba, F., and L. Orsenigo, *Technology Regimes and Firm Behaviour,* Oxford, U.K.: Oxford University Press, 1993, pp. 45–71.

[26] Malerba, F., *The Semiconductor Business,* University of Wisconsin Press, 1985, pp. 54–57.

[27] Vernon, R., "The Product Cycle Hypothesis in a New International Environment," *Harvard Business Review,* 1979.

[28] Clarke, C., et al., *British Electronics and Competition with Newly Industrialised Countries,* London: Overseas Development Institute, 1981.

[29] Steinmueller, W., "ICTs and the Possibilities of Leapfrogging by Developing Countries," *International Labor Review,* Vol. 140, 2001.

[30] Noam, E., *Telecommunications in Europe,* New York: Oxford University Press, 1992.

[31] Wu, T., *The Master Switch: The Rise and Fall of Information Empires,* New York: Random House, 2010.

[32] "Stripping Ma Bell," *Economist,* January 16, 1982.

[33] Christensen, C. M., "The Rigid Disk Drive Industry: A History of Commercial and Technological Turbulence," *The Harvard Business History Review,* 1993, pp. 531–588.

[34] Standage, T., *The Victorian Internet,* London: Weidenfeld and Nicolson, 1998.

[35] Young, P., *Power of Speech: A History of Standard Telephones and Cables, 1883–1983,* London: Allen & Unwin, 1983, p. 2.

[36] Manners, "50 Not Out (History of Transistor)," *Electronics Weekly,* December 17, 1997.

[37] Young, P., *Power of Speech: A History of Standard Telephones and Cables, 1883–1983,* London: Allen & Unwin, 1983, p. 22.

[38]   Young, P., *Power of Speech: A History of Standard Telephones and Cables, 1883–1983*, London: Allen & Unwin, 1983, pp. 36–40.

[39]   Moore, G. E., "Cramming More Components onto Integrated Circuits," *Electronics*, Vol. 35, No. 8, 1965.

[40]   "FCC Authorizes Fiber Optic Project," *New York Times*, May 25, 1984.

[41]   Young, P., *Power of Speech: A History of Standard Telephones and Cables, 1883–1983*, London: Allen & Unwin, 1983, pp. 91–107.

2

The U.K. Network

Now let's look at how the U.K. telecoms landline network developed. I will concentrate on the key period of 1950 to 2000 when major technology changes impacted this network (see Chapter 1). However, as will be seen, the events in this key period were significantly affected by decisions made much earlier and the ramifications continue to this day. In this period, there were also enormous changes in the structure and role of the network operator, which started in 1950 as an office of the U.K. Crown, generally referred to as the General Post Office (GPO) and ended as a private company BT listed on the London Stock Exchange; this is covered in Section 2.2. The key issue for the GPO and the supply industry was in the development and introduction of digital switching into the network. This development was eventually termed System X and came at a pivotal time in the history of U.K. telecoms; this development is discussed in Section 2.3. These areas of change were immensely challenging to the telecoms supply industry and I cover how they responded in Section 2.4. Here we see how the interaction of technical change, economic growth, the personalities involved, and government policy led to the virtual demise of the U.K. telecoms industry as a major source of telecoms supply.

The four sections are closely linked. To help view this, Table 2.1 gives the timeline of 1950 to 2000 split into the four sections with the key milestones mapped onto the timeline.

2.1   Evolution of the Network

The development of telephony has been chronicled in many excellent reviews, including [1]. However, the simple concepts of transmission and switching de-

47

Table 2.1

Timeline of Key Developments in the U.K. Telecoms Network

Year	Section 2.1 Network Evolution	Section 2.2 From GPO to BT	Section 2.3 System X	Section 2.4 U.K. Suppliers
1950		GPO Office of the Crown		5 suppliers to BSA
1955				AEI absorbs Woolwich
1960	Highgate Wood		Highgate Wood	
1961		Post Office Act	AGSD set up	Plessey acquires ATE and BET
1966	PCM transmission			GEC acquires AEI
1967	First Crossbar order			Closure of Woolwich
1968	Highgate Wood ended		Highgate Wood ended	GEC acquires Marconi
1969	End of BSA	Post Office Act		
1973	TXE4A rollout		TXE4A rollout	
1976			First System X order	STC leaves System X
1977		Carter Committee		
1981	Last Strowger delivery	C&W privatization	First System X delivery	
1984		BT privatization		
1985	Last manual exchange		BT orders AXE	GEC takeover of Plessey was blocked
1987			System X operational	GPT formed
1988	Start of ISDN			
1989				GEC takeover of Plessey
1990				Nortel takeover STC
1995	Strowger end			
1999		C&W U.K. sold		Full Marconi (GEC) take over GPT
2000				Marconi acquires RelTec and Fore

scribed in the previous chapter had been progressively developed and refined by the PTTs of all major countries. The need for automation of calls and functions such as billing systems led to increasingly sophisticated telephone networks. Figure 2.1 shows a simplified diagram of a local telephone exchange.

The U.K. telephone network had developed under the control of the GPO in the first half of the twentieth century using Strowger electromechanical

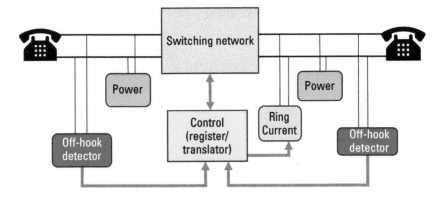

Figure 2.1 Functions of a local exchange.

switching (Section 1.6). By 1950, it was clearly suffering from underinvestment [2] and was regarded as having low productivity compared with other countries [3]. The U.K. telephone network was dominated by the use of Strowger switching, which had been first introduced in the United Kingdom in the 1920s [4].

The introduction of transistors into the transmission network was initiated in the 1960s. The development of pulse code modulation (PCM) transmission systems had been initially patented in 1938 by Alec Reeves [5]. Reeves was a U.K. engineer working in France for ITT, a U.S.-owned multinational telecoms supplier. This development of an efficient way of digitizing analog signals (such as speech) was the basis for the development of digital telecommunications. Its widespread implementation in the world's telecoms systems had to wait for the availability of transistors. The importance of PCM transmission was recognized in many ways, on a postage stamp (Figure 2.2) and a blue plaque (Figure 2.3). PCM transmission was first ordered by the GPO in 1966 when a series of multiplexed, multichannel systems were introduced notably using 12 and 30 speech channels per wire with a consequent cost saving [5]. The importance of the invention of PCM was rather belatedly recognized, with a postage stamp issued in 1969 and a blue plaque recording Alec Reeves' role, working at STL, Harlow (the U.K. development arm of ITT).

However, the application of transistors to switching proved much harder to realize. The GPO had been a key leader in the development of early computers in World War II, these had been mainly applied to code-breaking. It was optimistic that it could come up with a switching solution using time division multiplexing (TDM) coupled with pulse amplitude modulation (PAM). Early trials on this were promising, and the GPO decided that the U.K. network would keep the Strowger system until a full electronic system was available. This, in turn, led to a full trial exchange at Highgate Wood in London starting in 1960 [4], as illustrated in Figure 2.4. This was opened with much fanfare in

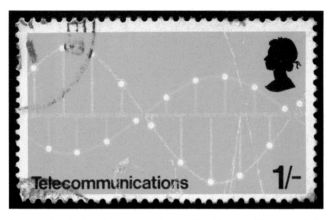

Figure 2.2 Stamp; nice design, but no mention of Alec Reeves. (Source: Wallace Henning, Shutterstock.)

Figure 2.3 Alec Reeves plaque. (Source: Harlow Civic Society.)

the early 1960s and claimed to be the first electronic exchange in the world. It was not strictly a digital exchange because the signals (sampled by TDM) were not digitally encoded. The exchange was developed by the GPO team under Tommy Flowers, who led the development of the wartime Colossus machine, an early computer used for code-breaking. However, the exchange's design (particularly the use of unencoded PAM) proved unsuitable for commercial-scale applications and the world went for the more sophisticated form of coding using PCM [6].

With the failure of the Highgate trial, a serious gap was opened in the country's development plan for updating its network. Only low-level development had been carried out on interim switching solutions, notably Crossbar, which had been developed privately by Plessey at Liverpool, and the reed-relay switching systems, which were initiated as GPO-funded developments, involving Plessey and STC.

Figure 2.4   Inside the Highgate Wood Exchange. (Source: BT Archives.)

By 1970, the world's telecom supply industry had broadly converged on how major countries' telephone networks would be digitalized. Let's start by looking at a simple schematic of the main constituents of a country's network, shown in Figure 2.5.

The structure largely derives from how the country's telephone network evolved [4]. Typically, local areas were first connected with later provision for long distance calls.

As can be seen, subscribers are usually linked to a local exchange (called an End Office in the United States). In turn, these exchanges concentrate the signals and route them on to trunk or international exchanges. In the United Kingdom in 2000, there were around 770 local exchanges and 76 trunk exchanges and relatively few international exchanges.

By the 1970s, there was a broad consensus that all voice signals would eventually be transmitted in digitally encoded PCM format. With the growing availability of semiconductors, digitalization of the telephone network could develop through a number of stages:

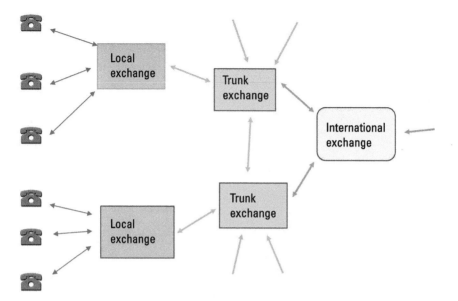

Figure 2.5 Simple telephone network schematic.

1. Digitalization of high-volume signal paths between trunk exchanges. This required relatively simple PCM encoding and decoding, multiplexing onto the transmission lines.

2. Digitalization of signals between local exchanges and trunk exchanges; here the signals could be digitalized before leaving the local exchange and decoded to analog signals only when arriving at the receiver's local exchange. In this case, trunk exchanges would simply only switch PCM-encoded signals.

3. Digitalization of local exchanges; here the signal from every subscriber line would be PCM-encoded either at the subscriber premises or at the exchange. This was a major challenge for developers of digital switching.

When the adoption of semiconductor switching proved much more difficult than the GPO had envisaged, progress was needed on the overall call-management of the switching network. Strowger was not well equipped to provide long-distance information on the routing of signals and some form of stored control of the ultimate destination of a call was needed. This involved a considerable amount of very clever development that culminated in the United Kingdom in the adoption of what was called subscriber trunk dialing (STD). Initial STD systems predated the transistor and used wired relay logic to provide what are called the register-translators in Figure 2.1. Electrical relays would be wired to store the necessary information used to control the electromechanical switches used by the exchange. Later versions used the components of current computer technology such as ferrite-core systems until semiconductor-based systems became economically viable.

To fill in the delay before a fully electronic exchange was available, the GPO turned to the use of reed-relay, hybrid switching systems that would use electronic control but reed relays (a miniature electromechanical switch) for the actual switching. This could achieve some of the advantages of semiconductor switching systems without solving the switching problems that were still to be overcome. The United Kingdom was reasonably successful in rolling out an electromechanical system under the names TXE2 and TXE4. These were developed for local exchange applications. Many other world suppliers of telecoms produced similar products, and, by the time the U.K. exchanges were introduced, other worldwide suppliers of telecoms had similar and well-established products. In some cases, these were introduced considerably earlier (ITT 1967 and L. M. Ericsson in 1968 compared with the U.K. TXE4 entering service around 1972). These products were also more sophisticated than the U.K. systems, for example, using stored program control (where the switching of the exchange is controlled by a stored program) rather than the TXE4's hardwired control. The United Kingdom had to wait for the later introduction of TXE4A (an advanced version of TXE4) before it had a stored program control exchange.

Behind this top-level development would also go the requirement for a substantial reengineering of the whole telephone network for digital transmission and management, which was implemented by the GPO in the 1970s. The task of carrying through the necessary changes to facilitate digital operation of a network of the scale and complexity of the U.K. telephone network without causing major disruptions to users required a great deal of ingenuity and management.

The architecture required to implement the integrated switched digital network (ISDN) is well covered elsewhere [7], but to summarize some of the issues:

- Setting up long-distance calls, particularly into other countries' networks, requires information on the call destination to be stored independently from the series of pulses generated by the pulse dialing system used by Strowger exchanges. By this point, many other countries (notably the United States) were using a tone dialing system. The call information is held in what is usually called the Register.

- Similarly charging information for the call made needs to be stored and forwarded to the network's billing system.

- Many facilities that we take for granted today require a degree of intelligence in the network. These would include call-forwarding, diversion, and answer-machine function.

- A key part of ISDN is that the digitalization of speech is carried out at the subscriber's premises. This led the way to widespread digital connection of particularly business customers (see Chapter 4).

Engineering these functions into a phone network that included several generations of exchange equipment (the last manual exchange was not removed until 1985) and the need for continuous operation was challenging [8]. The exchange functionality described above is usually referred to as electronic program control and the electronic exchanges that were developed would use software to achieve these functions (often termed stored program control). However, during the implementation period of electronic exchanges (a period of over 20 years), interim solutions were required, sometimes using a hardwired logic that was less flexible than a software solution. The U.K. situation with its predominantly Strowger network was considerably more complex than some of its European counterparts with Crossbar networks so throughout the 1970s a considerable proportion of the U.K. telecoms engineering resources was taken up in supporting 3 different switching systems: Strowger, Crossbar, and reed relay (TXE4).

Virtually all major telecoms suppliers began to develop electronic network switching starting in the 1960s. Many made an intermediate hybrid development by introducing an electronic program control first to Crossbar switches but later using the reed-relay matrix. The key immediate benefit was not a reduced cost of network switching since the electronic components were then relatively expensive. However, by reducing electromechanical components much higher reliability was achieved, which greatly improved operating costs. Other secondary benefits became significant, such as the size of building required for a given telephone exchange being greatly reduced, freeing up real estate.

With hybrid switches being deployed from the 1970s, attention was turned to a fully electronic switch solution. The general acceptance of PCM as the format for digitalizing signals helped to make this task feasible. The GPO had pioneered the use of PCM digitalization in transmitting signals using coaxial copper cables but was late in upgrading its switching by introducing Crossbar, hybrid switches, and a full electronic switching system. This will be analyzed further in the next sections.

The U.K. network was relatively quick to adopt the use of fiber optic transmission as it had been largely developed in the United Kingdom, with a trial cable laid by STC in 1976 [9]. The United Kingdom was in a strong position to develop fiber-optic undersea cables and the first transatlantic fiber optic cable (TAT-8) went into operation in 1988 [10], having been supplied by an international consortium that included BT providing cables from STC.

With the privatization of BT, the way was open for the development of digital services to take advantage of the growth in transmission capability. There was a dramatic growth in the use of fax (facsimile) machines (Section 4.4) and other digital services were developed using modems (Section 4.3). While facsimile transmission was first invented in the 1800s, it required the development of international standards and the availability of low-cost printers for the market to take off [11]. Xerox took the lead with the LDX unit patented in 1964; however, the industry only grew rapidly when plain paper printers using technology developed from photocopiers became available. Japanese manufacturers entered the market in the mid-1970s having already established a strong business in their domestic market.

The introduction of ISDN from 1988 when the standard was defined by the CCITT accelerated the process of data communications with a dual 64-kb connection becoming available as standard through the U.K. network (Section 4.6). The standard provided a path by which a network that had been conceived only to provide speech connection could now provide interconnection for both speech and data transmission.

2.2   From GPO to BT

In 1950, U.K. telecoms was seen as a mixture of telephony and telegraphy and operated as an adjunct of the U.K. PTT, the GPO. The GPO dated back to Victorian times when it introduced the first national postal system with a fixed price stamp to send a letter anywhere in the United Kingdom (the famous "penny black"). Because of this history, the GPO operated as an Office of the Crown, which, in practice, gave it enormous monopoly powers and immunity from much legislation controlling business. It reported to the Postmaster General, who was a government minister.

The United Kingdom's strong technological base included excellent abilities in many communications technologies. The GPO had strong technical capabilities (early computers were designed at the GPO research center at Dollis Hill, London, during World War II). However, the telecoms industry (from a user point of view) was weak, particularly with regard to the telephone service.

The United Kingdom had always lagged the United States in the production and rollout of telephone equipment. In 1914, there were 1.7 telephones per 100 population in the United Kingdom compared with 9.7 in the United States. In 1950, there were 5 million telephone lines installed in the United Kingdom (still only half the numbers per head in the United States), but there was a waiting list for over 500,000 lines.

The reasons for the weakness of the telecoms network included the long-established and efficient postal service, which lessened the potential demand for telephony. In 1900, there were 4 collections of letters each day in London, and it was possible to post a letter that would be received only a few hours later. Similarly, a global network of telegraph cables was established in the 1800s, largely to support the British Empire. This was regarded as adequate for most communications, whereas, in the United States, the telephone was seen as a vital part of municipal infrastructure to have a strong functioning local telephone network.

As in most developed countries, the GPO was a state monopoly that saw little point in addressing the full market opportunity. Initially, the Post Office had largely ignored the early development of the telephone system, which was left to private companies to develop. However, in the early 1900s, these began to impact the Post Office's monopoly of postal services and were nationalized in 1912 (with the strange anomaly of the city of Kingston-Upon-Hull, which maintained a municipally owned independent telephony company).

Until 1961, the GPO operating under the Postmaster General was tightly controlled by the U.K. Treasury and was not able to secure the kind of funds necessary to greatly expand the service. Thus, the United Kingdom (and Europe in general) was poorly supplied with telephone lines compared with the United States.

This issue had been recognized as early as the 1930s when the Bridgeman Committee Report [12] was published in 1932. This led to some improvement, but failed to address the central issue of a network that was unable to raise adequate capital to expand and improve its services.

On the technology side, there was not much innovation until the 1960s. Automation of the telephony service was under way but used the U.S.-invented Strowger system. In 1950, there were five major suppliers to the U.K. telecoms industry, which operated the Bulk Supply Agreement[1] as a cozy cartel managed by the GPO. Although the GPO had established excellent technological credentials by being one of the pioneers of computer technology during World

---

1. Bulk Supply Agreement: This was the main supply arrangement between the GPO and its suppliers. This arrangement was put in place in 1933 and was to run until 1969. The overall agreement covered most items used in telecoms including handsets, cables, and exchange equipment. Supply was only allowed by eight approved suppliers that were all members of an organization called Telephone Equipment Manufacturers Association (TEMA). For exchange equipment, supply was initially split equally between the five suppliers (using their original names): AEI/Siemens (Woolwich), British Ericsson (Nottingham), AT&E (Liverpool), GEC (Coventry), and STC (Southgate). Under this arrangement, pricing of items supplied were set by the GPO using audited data from two sample factories (Southgate and Liverpool). Pricing was set on the agreed costs plus accepted profit margin. With the takeovers of AEI by GEC in 1968 and AT&E and British Ericsson by Plessey in 1961, the agreement continued to run with GEC and Plessey getting 40% of the supply each and STC getting 20%.

War II, this came coupled with a belief that they would be capable of developing virtually on their own all the technology needed to modernize their telecoms system. This was consistent with the relationship with the supply industry where the key features of the telecoms network were developed by the GPO using the telecom suppliers as junior partners in development who subsequently were to supply the equipment developed under the Bulk Supply Agreement, which provided little scope for product innovation or development.

Unfortunately, a lot of the GPO thinking was based on a method of switching speech that proved to be impracticable and, as previously described, the failure of their early development of electronic switching with the trial exchange at Highgate Wood in 1960 left a major hole in the development of the network.

The supply side of the industry consisted of the 5 companies (AEI/Siemens, British Ericsson, AT&E, GEC, and STC) that were long-established makers of electromechanical switches and related equipment. Contracts were awarded by the GPO in relatively small amounts aimed to keep all the suppliers in business with little thought about cost reduction or industry rationalization.

The demand for telephones greatly increased in the 1960s and there was a great deal of political discontent with the slow rate of installing the telephone lines. The suppliers came under pressure to expand output and mainly responded by opening satellite factories rather than expand their main factories. Most of these satellite factories were sited in areas where the government offered financial incentives to locate them. These were depressed areas of large unemployment, notably Scotland, South Wales, Merseyside, and Northeast England.

Thus, the industry entered the 1960s without seriously addressing the issues that were to plague it. The industry followed the GPO's lead to remain exclusively supplying the obsolescent Strowger switching systems with a highly fragmented supply base and a weakening position in overseas markets. The declining competitiveness in export markets led to some privately funded development work being undertaken into Crossbar electromechanical switching, notable at AT&E Liverpool and AEI Woolwich. The Liverpool-developed Crossbar exchange was later adopted by the GPO.

By 1961, it was clear that the GPO had to be gradually separated from the government department in order to set a framework where it could operate more efficiently. The Post Office Act of 1961 gave a greater separation of finances, which was a precursor to the privatization that happened later. At the same time, an advisory group was set up by the Post Office to look at how to develop the systems. It was termed the Advisory Group on Systems Development (AGSD) and had a wide-ranging brief to steer the country towards electronic switching and all that went with it. This was a challenging brief because the Strowger system was a poor starting point compared with the Crossbar system used throughout most of the rest of the world. Considerable technological development was required to make the strategy work with STD, which was needed to enable calls to be placed automatically through automatic telephone exchanges. The program to introduce STD throughout the U.K. network started in 1958 and was not complete until the 1970s.

In 1969, there was another Post Office Act that readied the Post Office for later privatization with separate management for telecoms and post services. That year also saw the end of the Bulk Supply Agreement by a reluctantly commercial GPO. At last, there could be competitive supply arrangements although in practice there was not a dramatic change in the nature of the supply for another 10 years [13]. The original 5 suppliers of telephone exchange and transmission equipment in 1950 had, by a process of mergers and acquisitions, become the 3 key suppliers for 1970 to 1989: GEC, Plessey, and STC.

With the failure of its Highgate Woods trial installation, the GPO saw that the development of a fully electronic exchange was some distance away and interim systems were needed. From the late 1960s, the Post Office and its suppliers faced a turbulent time. The demand for telephone lines was rapidly increasing and there was considerable political pressure on the industry's failure to address the shortfall in supply. The ending of the Bulk Supply Agreement in 1969 left the industry struggling to establish an appropriate way of working together. At the same time, the network remained largely dependent on the obsolescent Strowger switching system, with the suppliers struggling to ramp up production from their satellite factories. Even in 1975, Strowger exchanges still accounted for over 30% of new exchange orders [14].

A privately developed (Plessey) Crossbar exchange was deployed starting in 1968. An interim development of a small exchange based on reed relay switching with electronics control had been carried through by Plessey and eventually released as TXE2. In turn, STC had taken up this development to create a larger exchange termed TXE4. Both were ordered from the early 1970s to fill in the perceived gap before the availability of a full electronic exchange but were not available for volume deployment until well into the 1970s. Even these interim solutions of a Crossbar and reed relay exchanges struggled because the Post Office was not able to produce a unified strategy for the development of a stored program control system for these exchanges so they had to rely on development-intensive work to integrate them into the network. The absence of the SPC also seriously damaged the export potential of these products. These and several other issues were pointed out in a long letter from the then-chairman of Plessey, John Clark, to the chairman of the Post Office, Bill Ryland, in December 1972 [15].

Onward development of the TXE4 exchange was carried out by STC to convert it to electronic stored program control (earlier versions used ferrite cores for data storage). This was introduced under the name TXE4A and would form a key part of the divorce settlement for the STC when it was eventually carved out of the development of the fully electronic exchange. By the 1970s, the PCM modulation system had been widely adopted throughout the world including the United Kingdom as a transmission standard for speech and this was rolled out by the GPO with some success.

The 1970s proved to be a period of dramatic swings in the politics of the United Kingdom. It started with the Wilson-led Labour government, which was intent on developing a strong industrial base for the country by revitalization of some industries (such as steelmaking) and encouraging new industries (such as nuclear power). This was followed in 1973 by the Conservative Heath government that was committed to the free market and took the United Kingdom into the then-European Common Market. A major jump in oil prices triggered by war in the Middle East caused high levels on inflation in the mid-1970s, which, in turn, put a major strain on government finances. The Labour government from 1976 under Callaghan remained committed to an interventionist industrial strategy but was severely constrained by its weak finances.

In the United Kingdom, the Post Office Corporation had been created by the Post Office Act of 1969 and had a clear role to dramatically increase the provision and quality of telecoms services, such as exchange lines. The resulting growth is shown in Figure 2.6, which is extracted from United Nations (UN) and GPO data plus that quoted by Harper [16].

In the industry, exchange lines are used as a key metric of capacity. It should be noted that each line required not just a physical connection to a phone but the provision of corresponding switching capacity at a local exchange. While the net installation of lines in the United Kingdom had been increasing from 1965, it rapidly grew in the 1970s. However, the chart also highlights the dip in line installation around 1975, which was caused by financial constraints on the GPO.

2.2.1 Privatization of BT

The privatization of the telecoms interests of the GPO to become BT was a significant milestone in the history of the U.K. telecoms industry. While much of

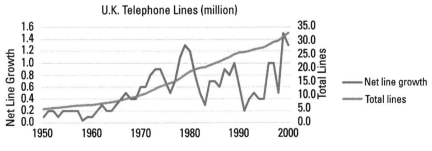

Figure 2.6 U.K. telephone lines.

the groundwork for this privatization had been carried out over the preceding 20 years, it still was a major shock for the supply industry (Figure 2.7). Many publications have been written on the subject, one of which by Mark Thatcher (no relation) compares the approach of the United Kingdom and France to privatization and regulation [17]. Interestingly, although radically different approaches were initially adopted, the two countries largely converged on the same long-term arrangements for running their telecoms networks.

The issue of privatization is itself still widely debated and will no doubt continue to be controversial as long as there is political debate in the United Kingdom. The issues largely fall outside the scope of this work and have been widely explored from many viewpoints. A Conservative party view was expressed in a paper published by a think-tank, the Institute for Government [18].

Many observers see the privatization of what had become BT in 1984 as a significant milestone in the development of the British economy and a ringing endorsement for Thatcherism. In practice, it was more complicated than that. The privatization was a natural development of a process that has been going on

Figure 2.7 Buzby character introduced in 1976 helped establish a new, livelier image for BT. (Source: BT Archives.)

for decades relating to the progressive separation of the telecoms network from the direct control of government. Key steps in this were:

- The 1961 Post Office Act, which separated the finances of the Post Office and removed the role of Postmaster General and the direct reporting line to a minister.

- The 1969 Post Office Act, which created The Post Office Corporation subject to many (but not all) of the rules of a normal commercial organization, rather than operating with the privileges of a Crown Department (and hence being immune to much legislative control). In particular, the telecommunication business and the post activities were clearly separated, enabling the subsequent creation of BT to what was to become a listed company.

- From 1975 onwards, the telecoms business was increasingly being run like a normal company with the improvement of budgetary control (accountants were virtually absent from the organization in the 1960s) and clarification of the organization and governance structure. It was subjected to serious review by the Carter Committee [19] of 1977 who concluded that the performance of the business had improved but was still not equal to that of some other major countries. In 1977, the entire GPO operation (including post) employed 400,000 people but only 54 qualified accountants.

- The British Telecommunications Act of 1981 formally transferred the telecoms assets of the GPO to a public corporation British Telecommunications.

The privatization of BT was well supported by small and large investors alike who were able to see that a business with a virtual monopoly in a rapidly growing market sector was likely to be profitable almost regardless of the competence of its management. The actual privatization process and its results have been widely covered in other publications and will not be greatly explored here; however, a few points should be made.

First, there is a widely held belief that the privatization of BT itself was a catalyst that led to a dramatic improvement of the productivity of the telecoms sector. However, many have argued (notably, Harper in several publications) that the growth in efficiency actually started well before privatization took place and as shown by the installation of lines shown in the graph in Figure 2.6. The growth correlates strongly with the implementation of electronic switching and transmission rather than any link to privatization. Clearly privatization enabled the newly formed BT to raise adequate funding at a time when it was needing to heavily invest in its network. This contrasts with earlier periods in the growth of the telecoms network when the shortage of government funding meant that investment was restricted to the detriment of the supply industry, notably in the period of 1975 to 1979. The spectacular growth in operational productivity came from the dramatic reduction in support costs arising from the electronic equipment that enabled BT to steadily reduce its manpower from the 1980s. Thus, in the 25 years from 1970 to 1995, the number of telephone lines increased by 150%, but the staff numbers reduced by almost 50% despite steadily increasing revenue [20]. This meant it was inevitable that the newly formed BT would, for many years, show improving financial performance to the delight of its investors.

A second point is that there was a perception at the time of privatization that BT would be controlled by market factors as competition entered the market and this would ensure that BT operated in a fair and efficient way. However, as an interim step, a regulator, Oftel, was set up in 1984. The expectation has been dramatically missed over the following decades mainly because of the enormous cost of building a parallel network to compete with the BT network. Oftel [later renamed Office of Communications (Ofcom)] has become part of the fabric of the U.K. telecoms industry. The capital-intensive nature of landline telephone networks makes parallel-wired networks unviable. As will be seen in Chapters 3 and 4, the advent of mobile telephony and digital communications would introduce indirect competition to fixed-line telephony and this would include some degree of sharing of the basic infrastructure, but the direct competition envisaged at the time of BT's privatization never stood a realistic chance. This is explored later.

The newly privatized BT was clearly a flagship of the government of Margaret Thatcher and was anxious to follow policies in line with government thinking, maintaining a significant investment in the U.K. telecoms network while delivering strong financial performance. One of the prices of this situation is that BT was encouraged to largely abandon its interests in making and supplying equipment for its network (much to the disgust of some of its employees); some very successful businesses (such as in fiber optics) were sold off and to a considerable extent its research and development capacity were reduced. Its flagship research site at Martlesham, Suffolk, is now largely an industry park.

As will be discussed later, the rest of the U.K. telecoms supply industry struggled to deal with the consequences of the dramatic change in thinking in two key areas: the inevitable need to reduce manpower and the simultaneous need to build for the future around the successful implementation of electronic switching embodied in the U.K. System X development. One year after privatization, BT ordered in parallel what it described as the System Y electronic exchange, which was based on the Swedish AXE switch.

2.2.2 Competitors to BT

An implicit part of the case for privatization of BT was the belief that it was possible to use competition in the marketplace to eventually replace the need for state-controlled regulation. As described earlier, this proved to be an over-optimistic belief in the power of unregulated capitalism. The U.K. government actively engaged in a long-term attempt to generate competition for the monopolistic BT. This can be described as three separate phases:

1. *The creation of a privileged competitor, namely Mercury Communications.* This was created before BT was privatized using the resources of Cable and Wireless Ltd. Cable and Wireless was a remarkable enterprise being an assembly of colonial era telecoms interests that was particularly strong in providing international communications, notably to Hong Kong and Bahrain. Its roots lay in the Victorian beginnings of underseas telegraphy cables when it operated under the unusual name of The Gutta-Percha Company; Gutta-Percha is a resin from a Malaysian family of trees whose gum was suitable for providing waterproof casing for cables. While it had been a rather low-profile operation with pedestrian results up to the 1980s, efforts were made to liven it up and the company had been privatized in 1981 as a precursor to BT's privatization [20].

    At around the time of its privatization, Cable and Wireless set up what was to become Mercury Communications with top rank partners, British Petroleum and Barclays Merchant Bank (both of which could be influenced by the U.K. government). Mercury's objective was to attract U.K. business customers, particularly those wanting international communications, since Cable and Wireless had strong links to international cable operations. To support the required investment, Oftel granted Mercury a unique 25-year renewable license to compete with BT and they were guaranteed to be the only competitor until 1990 (Figure 2.8). Mercury planned to put together a network of fiber cables serving major population centers with an investment approaching £1 billion, but this proved to be too rich for Cable and Wireless's partners who bailed out of the consortium. Despite this investment, most customers had to use BT for linking into this network. Plans to build up trade by undercutting BT were largely scuppered by BT reducing its international prices but not being so generous in pricing its local connections where it mostly enjoyed a monopoly.

Figure 2.8 Mercury phone box. No competition to the iconic red phone box. (Source: Consallforge CC-SA 4.0.)

The venture went through many twists and turns to attempt to reach serious sales volumes; this included linking up with Electrical Power Distribution Companies to provide an easier way to provide a fiber optic framework for its network, using their pylon network. It also promoted a short-cut way for subscribers to enter its network by using an access code or hitting the "blue button" on its specially supplied phones. All this proved to be inadequate to make any significant penetration against BT's entrenched position.

In the 1990s, Mercury was increasingly involved with cable companies setting up in the United Kingdom and provided over 100,000 phone lines for cable customers. In 1996, the business was merged with 3 other cable companies (Vidéotron, Nynex, and Bell Cable media) and renamed Cable & Wireless Communications. In turn, Cable & Wireless's stake in this U.K. consumer business was abandoned in 1999 with the sale of its residual U.K. operations to NTL (later to become Virgin Media). Cable & Wireless itself was eventually taken over by a new Chinese telecoms conglomerate, Pacific Century Cyber Works (PCCW) in 2000. In what was to become a common pattern, the long-established player was merged with the newcomer with mysterious funding sources and then descended into chaos and oblivion.

2. *The entry of new-media competition.* The growth of satellite and cable TV in the United States had been noted by the U.K. government and it was felt that the U.K. TV market, which was dominated by the publicly owned by Royal Charter British Broadcasting Corporation (BBC), could do with some competition. Thus, licenses were awarded to both suppliers of satellite TV and cable TV franchises, who were

expected to supply new media content to compete with the BBC and its terrestrial broadcasting competitor ITV.

The commercial basis of these newcomers was to provide a subscription-based service offering premium products, particularly sport. The commercial operators, notably Sky with its satellite service, were able to outbid (and bid-up the price of) sports events and hence gain a significant foothold in the market. However, cable operators were not able to operate so effectively as they had to systematically lay out a cable network and could not respond quickly to peaks in demand. U.S. experience had led them to expect a take-up of over 20% to their service, but because of the quality of the terrestrial broadcasting and the nimbleness of the satellite operators, they achieved nothing like those figures; even by 2020, the penetration of cable TV in the United Kingdom was only 12% compared with 44% in the United States (data from statista.com). The cable operators eventually consolidated into one operation operating under the brand Virgin Media, which was eventually bought out by Liberty Media of the United States in 2013 [21].

However, a substantial part of the country had been cabled and could in theory provide a competitive alternative to BT's monopoly of the last-mile issue (often referred to as the local loop). Unfortunately, the topography of a cable broadcast system is very different to that of a telecom network and the merged cable suppliers had neither the resources nor the inclination to invest the extra billions of pounds to turn their cable network into a serious competitor to BT, although this remains a remote long-term possibility.

3. *Network sharing with broadband suppliers.* This has proved to be the most durable approach to limit BT's monopoly power. Independent service providers can resell connectivity on the BT network as part of their own branded offering; thus, internet service providers (ISPs) who provide internet connectivity can use BT's network to connect their customer base to their servers. This arrangement is perhaps the simplest and has enabled BT's headline share of the domestic market to decline steadily as third-party suppliers of internet connectivity (such as TalkTalk and Virgin Media) can offer bundled services including voice telephony and broadband.

There are several ways in which network sharing has been used. Perhaps the simplest is that the BT network is often used for broadband connectivity between the data centers of ISPs. In addition, connectivity to end users is often provided over the BT network, initially through their phone lines using modems, but more recently using asymmetric digital subscriber line (ADSL).[2] Thus, the charges to consumers hide the payments the ISPs must make to BT to provide the connectivity, which, in turn, leads to regular debates with Oftel/Ofcom. The picture is further confused by BT offering its own bundled service of voice call and broadband. It even introduced an independent supplier, PlusNet, which is owned by BT but operates independently at lower prices, thus squeezing the margins of other ISPs in a way similar to how BT squeezed the margins of Mercury for international calls in the 1980s by reducing its own tariffs.

In Chapter 4, I explore the development of data communications as it evolved from simple signaling over voice connections through the development of ISDN through to ADSL and full fiber communications. To date, the commercial policies of BT have largely controlled how these services have evolved since even apparently independent U.K. telephone network operators are largely dependent on buying communications from BT. Thus, take-up of early data communications on voice and ISDN lines was quite slow because users paid for connectivity by the minute (as with voice calls) and BT's pricing was not very attractive to domestic consumers. The introduction of ADSL [22] led to the offering of a fixed daily fee for connectivity, which proved to be much more attractive to users and hence accelerated the growth of data traffic.

The requirement of high bandwidths to enable the transmission of video images to consumers was expected to make the BT copper-based network obsolete. However, this was effectively countered by BT widely introducing ADSL transmission over its network starting in 2000 with the bandwidth achieved by this growing to 10 Mbps and higher, which is good enough to transmit high-definition TV signals. Thus, BT, through its Openreach subsidiary (which had been created to run BT's U.K. network), could concentrate on building up the capacity of its main network while leaving most of the telephone subscribers in the United Kingdom still reliant on copper wire connection to a local connection point. Even if a rival network supplier wins a domestic account, it will invariably be forced to pay BT a rent for the use of the last-mile copper connection to the consumer. ADSL

---

2. ADSL was a major bonus to the owners of traditional copper-based telephone network. The network had been engineered for voice frequency use, up to about 4,000 Hz, and by 1998, modems were capable of carrying 28.8 kbps, sometimes faster, on many voice-grade calls. ISDN had by then stretched the loop plant to run at 160 kbps. ADSL, using more advanced signal processing, shared the loops with telephone services by running in frequencies above the voice range, and was able to extend the downstream capacity of many copper lines to as much as 10 Mbps, though with much lower upstream speed.

proved adequate for most consumers, although the growth of applications requiring a higher upload speed than ADSL typically provides (such as gaming and video calls) would increase the pressure for higher bandwidth links. The pressure on fiber to the home (FTTH) or even fiber to the kerb (FTTK) thus remained at a relatively low level in the United Kingdom until the 2020s when, under increasing pressure from customers, BT committed to rolling out a full-fiber network by 2025 [23].

4. *Future competition.* The wide availability of ADSL through the BT copper network reduced the pressure for the installation of other networks such as cable TV. It will need either strong political will or a market demand for bandwidth beyond the typical 10-Mb of ADSL to generate much momentum for FTTH. Thus, while in the longer term, the cable TV fiber optic network could offer serious competition to what is currently the BT Openreach operation, this remains unlikely as a full-service competitor, although the piecemeal sharing of network infrastructure (both high-capacity datacoms between operational centers and local loop) does provide a degree of competition.

In many ways, the only effective competition has been the growth of mobile phones (see Chapter 3). These have reduced the need for fixed voice telephony, and, in most developed countries, the number of fixed-line installations has stayed rather constant since 2000. It is the demand for broadband connectivity that has largely preserved the BT position. While the increasing bandwidth of mobile telephony makes it possible that fixed communication will become obsolete, this still seems unlikely even with the projected rollout of 5G mobile phone networks.

In the absence of serious competition, BT continued to be closely monitored by the U.K. government, which retained a "golden share" in the company preventing it from being taken over. It was regulated by arrangements such as that provided by Oftel (now named Ofcom), a U.K. regulatory organization. The regulator aimed to prevent BT from abusing its monopoly powers with regular threats to refer BT to the U.K. Monopolies and Mergers Commission. This threat was usually effective in gaining BT's agreement to various forms of price restraint. This was applied throughout the 1990s as a policy of pricing of "RPI minus X%" where overall prices were set to increase by an agreed number (typically 7.5%) less than the growth in the Retail Prices Index. The close control over BT sometimes caused problems, notably when BT sought to merge with the U.S. telecoms operator MCI in 1997 [24]. This led the government to abandon its golden share. That analysis also shows that, although applied in different ways, the regulation issues of the U.K. and French operators in the period of 1970 to 2000 were very similar. Indeed, the very dynamics of the national telephony networks made this consistent with experience with privatization in most other countries.

The hard truth for Thatcherites is that for almost all countries' utility networks (electricity, gas, water, and telecoms), the cost of creating the infrastructure prohibits the creation of widespread direct competition. Hence, countries have to choose between state ownership or state-controlled regulation if they are to avoid monopoly exploitation.

2.3   System X

We now come to a key moment in the story of the U.K. telecoms industry. In shorthand, it is usually referred to as System X, but this simple if enigmatic title covers an epic story involving several larger-than-life characters and the efforts of thousands of the country's best engineers as they wrestled with the issues of equipping the U.K. telecoms network for the twenty-first century.

First, let us look at the background. While the application of semiconductors to telecoms spans a wide period from the 1950s onwards, perhaps the most important period was the 1970s when most Tier 1 telecoms suppliers in the world started the development of electronic switching for the telecoms network. By then, it was clear that the economics of telecoms networks were shifting dramatically away from the electromechanical systems that had been used for over 50 years in favor of using digital technology based on the increasing availability of semiconductor integrated circuits and software-based systems.

2.3.1   Electronics in Telecoms

The concept of using electronics in telecoms networks was well established, even before the invention of the transistor (1947). Valve amplifiers had been used in transmission systems since the 1920s and the GPO, which had been heavily involved in development of early computer systems in World War II for applications such as code-breaking, was of the clear opinion that the country should make a direct step into electronics once the impact of the development of the transistor was understood.

While telecoms networks had been automated since the beginning of the 1900s, they remained essentially a system for providing a wire connection between the sender and the receiver. Amplification had been introduced in the transmission lines and the connections used multiplex technology adopted to reduce the amount of copper in the network. However, the network remained essentially simple and analog.

Benefitting from new electronic components required not just the availability of these components but also a total rethink of telecoms network architecture, particularly the adoption of digitalization of the analog speech signals the network handles. Early attempts at creating electronic exchanges were pioneered by the GPO, notably the experimental Highgate Wood Exchange, which was claimed in 1962 to be the world's first electronic exchange. However, it failed to achieve satisfactory performance in a real operational situation and was switched off in 1968.

With the introduction of the transistor, it was expected that this major improvement in technology would benefit telecoms; the costs of implementing the network would be reduced and the higher reliability of solid-state devices would improve reliability and hence operating costs. However, the automation of telephony required more than just an electronic switch. This is discussed by Valdar in his article on the history of switching [1]. There is the need for a register/director function that would record the details of a call setup to ensure that signals are routed correctly through the network and call functions (such as charging) are recorded. Telecoms manufacturers found the design of the register/director function was well suited for the application of digital electronics prior to applying it to switching signals.

A significant factor relating to the U.K. telecoms network was the high level of Strowger switching within the U.K. network, as discussed previously. This technology had been largely superseded by Crossbar in many other countries' networks. Crossbar-based networks already needed a significant register/director function, and this led the indigenous manufacturers into an evolutionary path in the development of electronic exchanges, which, as Valdar described, leads into intermediate switching systems using reed relays before a fully digital exchange is developed.

In the United Kingdom, as a result of the failure of the early PAM technology and the extensive use of Strowger, the development of an electronic exchange was both complex and late in starting. These issues would be recurring theme in the development of what became System X.

2.3.2 The Birth of System X

In the 1970s, electronic exchanges began to be introduced in various countries, but it was becoming clear that the United Kingdom was not well placed in this development. The failure of the trial electronic exchange in Highgate Wood in the 1960s was a considerable setback. An Advisory Group on Systems Development (AGSD) had been formed by the Post Office in 1961, but no major new commitment was made to developing an electronic telephone exchange.

The Post Office reluctantly started buying Crossbar electromechanical exchanges in 1968 when it was clear that no electronic switching solution was imminent. Its horror in finding itself in a sole supply situation led to its supplier, Plessey, being persuaded to license its 5005 Crossbar design to GEC as a condition of supply.

Other countries that were mainly using Crossbar switching had started using software-based controllers for exchanges, using stored program control.

Hybrid interim solutions for telecom switching were being widely adopted throughout the world but again the U.K. adoption of this was late compared with, for instance, AT&T in the United States, which started installing their hybrid 1ESS switch, first of a family of switches in 1965 [19]. A U.K. version had been developed by the Post Office in conjunction with Plessey and STC. Initially, this was a local exchange using a fixed control (rather than the more advanced stored program control) called TXE2. A large exchange version was then developed called TXE4, and this culminated in a stored program controlled version of the exchange called TXE4A being rolled out from 1973 by the STC. As Figure 2.9 shows, even in 1980 around 75% of all local exchanges in the United Kingdom were Strowger [25]. Note that the graph shows exchange modules, many of which are installed together in a building to form complete exchanges as shown in Figure 2.9.

The development of a fully electronic exchange system was delayed due to its early setback (above) and the complex way in which the development was envisaged with the Post Office leading the development with all three suppliers also participating. This contrasted poorly with most other developments of competing electronics switches throughout the world, which involved typically one major commercial supplier with the PTT acting as the initial launch customer. It was also clear that development funding would be needed, and the government was the only possible source of this at a time when industry policy had many other competing financial commitments.

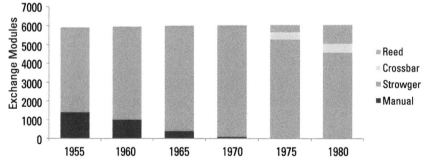

Figure 2.9   Exchanges installed in the United Kingdom.

While the Post Office became an independent corporation, it was still dependent on government for funding and overall policy. This had a noticeable effect in 1973 when following the worldwide oil shock, the government decided to cut back the GPO's investment plans despite there being a clear need for continuing investment. This, in turn, had a drastic effect on the supply industry with redundancies and factory closures [26].

It became clear that a major procurement of electronic exchanges would be required and that other countries such as Sweden and the United States were already well advanced in this process with their indigenous manufacturers Ericsson and AT&T well into developing such exchanges in the early 1970s [27]. A delegation from the Post Office had visited Ericsson in Sweden around 1972 and were given a view of their program for the introduction of the competing AXE system as summarized in Figure 2.10.

Thus, when the new Labour government took office in 1975, it found the future of the telecoms industry on its agenda. There were two interlinked issues:

1. Whether it made sense to develop an electronic telephone exchange when other countries were well ahead in development of competing systems.

2. Whether it was tenable to have the telecoms supply industry fragmented between 3 competing companies and if so, how could a large development be successfully achieved?

Thus, numerous reviews and discussions took place; notably, the Carter Committee reported on proposals for what had been named System X and in 1976 a Cabinet Committee [27] was created to advise on these issues.

The input of the major personalities involved should not be ignored. Although the Bulk Supply Agreement had officially ended in 1969, the mindset of the GPO (as it then was) and the three main suppliers of telecoms, GEC, Plessey, and STC, were still rooted in the era. In particular, there was considerable personal antagonism between Arnold Weinstock (the CEO of GEC) and John Clark (the CEO of Plessey) [28], although telecoms suppliers managed to work together with reasonable relationships at the operational level. The personalities and companies involved are explored in Chapter 5.

Documents accessed from the U.K. National Archives give a useful overview of the thinking at the time. While some players (notably the Trade Union Congress (TUC)) wanted to achieve a merger between all three major telecom suppliers [29], Plessey, GEC, and STC, this was deemed to be politically impossible. In particular, the strong position of Arnold Weinstock as CEO of GEC (then the United Kingdom's largest manufacturing company) meant that the government did not have the appetite for forcing through that change. A lower key approach was widely debated whereby STC would be eliminated.

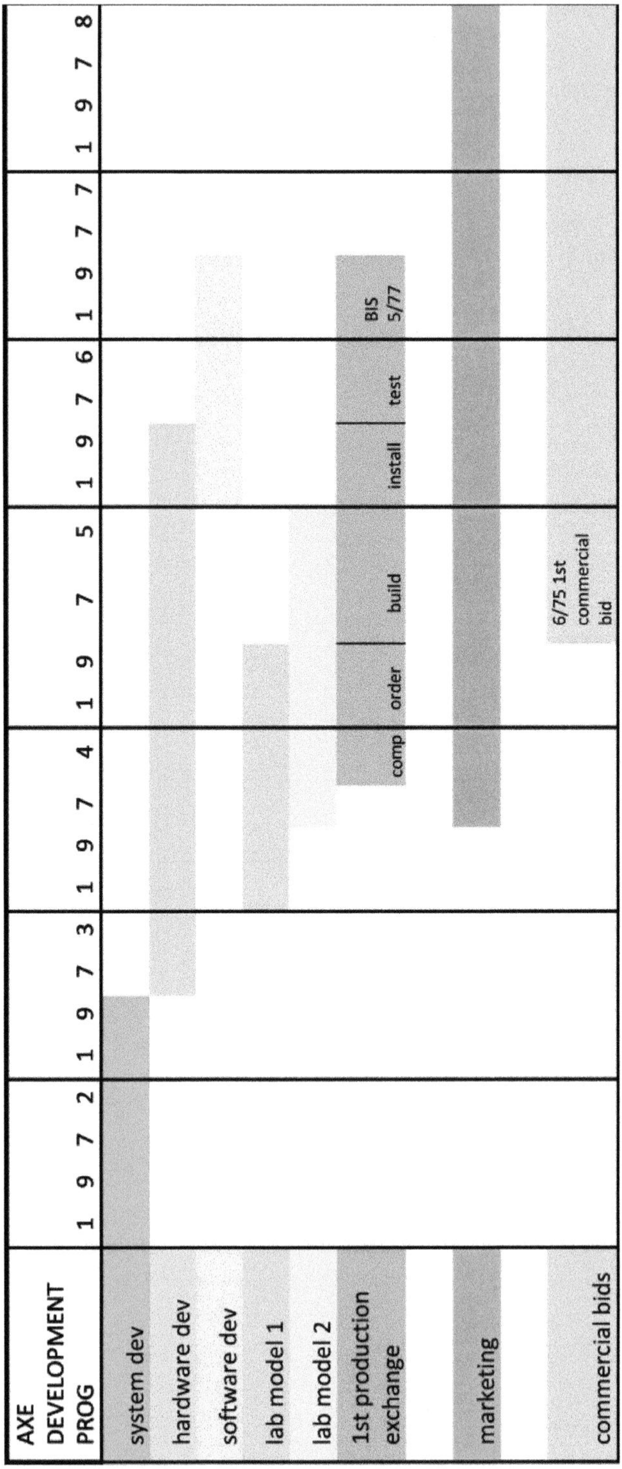

Figure 2.10 AXE development program.

The GPO's proposal to take over STC was rejected and a merger of STC with Plessey was agreed by the government to be the best way forward in 1978 [30]. The perception was that Plessey had technical strength but poor management, whereas STC had strong management but a weaker position. However, this decision ran into the political issues relating to the Winter of Discontent, which brought down the Labour Prime Minister Jim Callaghan's government in 1979 and consequently disappeared from the agenda.

This was a major lost opportunity. Both GEC and STC were vulnerable to government pressure at the time since they had been caught out on a major supplier cartel scandal relating to the supply of cables to the telecoms industry in 1975 [31]. In addition, the government had, unlike later years, a serious capability of promoting such mergers using the National Economic Development Organisation (NEDO). My general impression from the National Archives is that the government was busy discussing with GEC issues such as the running of the nuclear industry and had no bandwidth or appetite to use this in the context of telecoms. The TUC's paper in 1977 set out a very plausible blueprint for how the companies could be merged [29], but this was never followed up.

Many people who were knowledgeable about the situation and the industry (such as Sir Kenneth Berrill, head of the Cabinet Office) suggested the GPO should abandon the internal development of digital telecoms switching because they were too late and the development would prove to be delayed and unworkable [32]. However eventually these were overruled and the electronic system, which was by then called System X, was finally ordered by the GPO from a consortium of just GEC and Plessey. Initial orders were placed in 1976 with development ramping up over the following years. Funding for the project was provided by the government via the Post Office with a development cost envisaged at around £100 million [27]. Work was split between Plessey and GEC, with Plessey taking the lead role (but with considerable Post Office involvement).

No serious negotiations took place with outside providers of technology (such as Ericsson of Sweden) over a deal to license technology instead of in-house development. Clearly, the GPO maintained a close contact there, but it appears only as an alternative source of supply. One can only speculate what arrangements could have been made, but it would seem possible that a stronger U.K. telecoms industry would have emerged with a better chance of winning export orders.

STC was taken out of the equation and, as a consolation, was given main supplier arrangements for TXE4A [31]. This proved to a be a considerable cash-cow for the company as the inevitable drift in the System X project meant that there was a need for up to 1 million lines per year until System X came on stream for serious delivery in 1987 (around 5 years behind schedule) [33]. The successful development of System X was in itself a major achievement with collaboration between the GPO, Plessey, and GEC spread over more than 10 sites in the United Kingdom. However, the costs were well above the original estimate [34] and volume deliveries started late. The original planned times-cales were unrealistic being similar to those set out above by Ericsson for one company developing a product. The net effect of the delay in System X and the deliveries of TXE4A (here called Reed Systems) can be seen in the chart in Figure 2.11, which was assembled from a number of sources including the BT Archives and Harper [3].

In practice, the System X project overran in time due to several factors:

- Time required by subcontractors to recruit engineering staff.

- Technical complexity of the project hindered by the distributed working arrangements.

- Decision to use in-house components (such as the GEC Mark II Bl processor) rather than buy-in best-of-breed devices.

Because of the political sensitivity of the development the project man-agement was extensively monitored both by the Post Office and by govern-ment, but the project came in well over the initial £100 million with the final costs reported at £500 million (at least £350 million being spent by BT with Plessey, GEC, and STC) [35].

However, the real cost of the delay was the consequent failure to establish volume sales, which would support the supply companies. As can be seen in later chapters, their inadequate profitability led to their increasingly desperate search for mergers and alternative revenue streams.

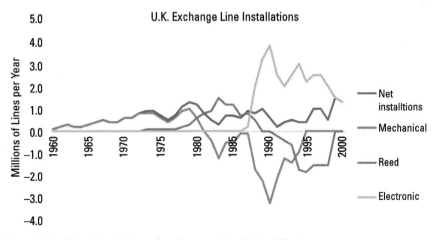

Figure 2.11   Yearly installation of exchanges in the United Kingdom.

A major strand of government concern at the time was to try to achieve export sales of System X. It saw that the United Kingdom's export performance in telecoms was declining and it needed System X to succeed to reverse this trend. An internal government paper in [27] pointed out that the United Kingdom had declined from what it claimed to be the world's largest exporter of telecommunications equipment in 1963 with export sales reported at £29 million to the fifth largest in 1975 when exports were reported at £66 million. However, while the United Kingdom did have a positive balance of trade in telecoms equipment in the 1970s, this was based on its dwindling control over commonwealth markets such as Australia where U.K. telecoms suppliers had long established subsidiaries. With the failure of the U.K. industry to provide technical leadership an increasing part of this market was being taken over by other suppliers, particularly Ericsson, which adopted a very successful second supplier strategy in many markets.

As discussed above, electromechanical switching started as simply a question of generating a wired path from sender to receiver. As networks became more sophisticated, they had to provide STD plus more secondary functions such as billing and call forwarding. This required the use of a much more complex infrastructure, which, in turn, required network-specific expertise. Increasingly selling telecoms required expertise on the entire network rather than just providing separate items of equipment. To achieve this an export corporation was set up to develop export business in 1978. Up to this time, the U.K. export efforts were fragmented since the GPO did not see itself as needing to be concerned with export issues. When selling telephone equipment began to require sophisticated network knowledge, the GPO was the prime holder of these skills in the United Kingdom and was not that bothered about using this resource to help exports. However, the nature of selling telephone exchanges required overall system expertise that largely resided with the GPO. In contrast, France was at the time regarded as an example of good practice where its PTT worked closely with its telecom suppliers to achieve an overall system capability.

The U.K. solution of a joint corporation, British Telecommunications Systems Ltd. (BTS), jointly owned by BT and the main suppliers (mainly GEC and Plessey), was eventually put together in 1979, but was riddled with internal issues and never really received the full-hearted support of its partners. The GPO had little interest in providing system application support to other countries and the antagonism between GEC and Plessey hindered effective cooperation [36]. Even if System X was technically and commercially competitive, it proved to be difficult to coordinate the sales process through BTS with the necessary development by subcontractors to make serious export sales (Figure 2.12). There is little sign that BTS made a serious impact on selling System X and most export successes came much later in the 1980s from what was by then GPT (GEC Plessey Telecoms). The failure of BTS put the U.K. industry at a

Figure 2.12   Nice display in Rio, but few sales. (Source: BT Archives.)

substantial disadvantage compared with the offerings by French and Swedish suppliers.

By the end of the 1970s, things had turned out much as people feared with the System X development running late and other countries rolling out competitive and cheaper electronic exchanges before System X could be credibly offered to export customers. Just what an uphill task was faced in the export market can be seen by Table 2.2 published in 1982, which gives the worldwide situation for electronic exchange lines.

Table 2.2
Ranking of Digital Telephone Exchange Suppliers, 1982 (Telephone Journal)

Company	Country	Exchange	Lines (Million)
CIT-Alcatel	France	E10 and E12	8.5
Northern Telecom	Canada	DMS	5.1
L. M. Ericsson	Sweden	AXE	4.3
NEC	Japan	NEAX61	3.8
Thomson-CSF	France	MT20 and MT25	3.6
Western Electric	United States	ESS	2.6
ITT	United States/Europe	System 12	1.1
Philips	Netherlands	PRX/D	1.1
GTE	United States	EAX	1.1
Siemens	Germany	EWSD	0.6
Stromberg-Carlson	United States	System Century	0.4
GEC/Plessey	United Kingdom	System X	0.2

In 1981, the last Strowger delivery and the first System X service delivery occurred, although in practice it was not until 1985 that the system went into serious use. The fears of many people that System X would prove to be expensive, overcomplicated, and difficult to sell outside the United Kingdom were largely proven to be true. At a later point in the System X history, a few notable export orders were received, one being from the Chinese railway system and another being from Colombia. This showed that System X was a viable product that in different circumstances could have been the basis of a successful export business [36].

Meanwhile, major changes in telecoms were taking place that bypassed the traditional U.K. telecoms supply industry. The creation of cellular mobile phone networks and the growth of data communications are explored in the following chapters. I would argue that a hidden cost of the System X development was that technical, financial, and managerial resources that could have been focused on these opportunities were focused on the System X development and thus contributed to the eventual demise of the United Kingdom as a major center for telecoms development.

2.4 U.K. Telecoms Supply

The supply industry entered the 1950s with five long-established suppliers running large integrated factories spread through England. These factories had mainly been retasked to making military equipment during World War II. Supply was split equally between these five suppliers. Interestingly, four of these companies originated from non-U.K. suppliers, but they would subsequently form the nucleus for the indigenous industry.

1. Automatic Telephone Manufacturing (ATM), later called AT&E, was based in Liverpool and had been well established to supply electromechanical switching, having been an associate of a Chicago company of the similar name (Automatic Electric) with rights to their Strowger product.

2. The Woolwich factory, usually referred to as Siemens Brothers. It was founded by two Siemens brothers in 1863, part of the family which also started the German Siemens conglomerate. From 1917, it operated as an independent company becoming a subsidiary of Associated Electrical Industries (AEI), a large U.K. conglomerate, in 1955.

3. The U.K. General Electric Company (GEC) had a subsidiary at Coventry, started in 1922, which supplied telecoms switching and transmission equipment. This could claim to be the only one of the five companies with no clear non-U.K. roots. Despite it having a similar

name to the U.S. General Electric Company (incorporated in 1892), its name dated back to a U.K. company founded in 1886.

4. At Beeston, Nottingham was the British Ericsson company (BET), which again had its early roots as part of the Swedish L. M. Ericsson company but was now operating independently after being bought out from Ericsson in 1948.

5. STC Southgate was part of the STC (Standard Telephone and Cables Ltd.) company, a subsidiary of the American International Telephone and Telegraph company (ITT). ITT was itself spun out of AT&T in the United States in 1925, acquiring most of the assets of the Bell business outside the United States when AT&T was concerned about antitrust legislation. Its country businesses, STC in the United Kingdom, LMT in France, Standard Elektric Lorenz (SEL) in Germany, and Bell Telephone in Belgium, operated almost completely autonomously.

It was generally accepted that five suppliers were too many and that the arrangements were inefficient with a large number of sites spread around the country (see Appendix B). This was not likely to encourage investment in the research and development that was going to be necessary to upgrade the network. Apart from Japan, most other countries had much simpler supply arrangements. Japan's situation and its development of an electronic exchange similar to System X shows that a fragmented supply situation does not in itself doom a country to failure. Apart from a more collaborative culture, the Japanese organization of research and development more effectively involved its suppliers. The relative weakness of the U.K. supply companies and the poor working relationships greatly weakened the U.K. situation.

In many ways, the GPO found the fragmented suppliers convenient as they wanted to maintain technological leadership on all developments. All these companies were supplying equipment based on the Strowger technology under the Bulk Supply Agreement and there was little encouragement to look beyond this as all big developments would come out of the GPO and it was widely expected that an electronic exchange would be developed by the Post Office.

The U.K. supply industry enjoyed a rather privileged position in export markets, particularly within the former empire territories (with the notable exception of Canada, which was dominated by U.S. originated suppliers). Telecoms was regarded as part of infrastructure investment and it was a natural consequence that British practice and suppliers were the first port of call when installing their fledgling systems. However, with the growth in independence from the original colonial power, coupled with the feeling that the United Kingdom was not a clear technical leader in this area, was an opening that other world suppliers (such as Ericsson of Sweden) were beginning to exploit. From the point of view of the telecom supply industry, the introduction of digital switching was a major disruption since their factories at the time were built to produce high volumes of small mechanical parts and organize them into the assembly of electromechanical switching systems. Usually, the factories were highly integrated without the need for extensive supply of external components. As such, these facilities were highly labor-intensive and required a substantial workforce with basic mechanical engineering skills.

Meanwhile, the supply industry had begun to see the need for rationalization and Plessey (a U.K.-listed company that was already a supplier to the telecoms industry) entered the field by acquiring in 1961 both AT&E and BET, which reduced the monopoly supply situation by one. Perhaps more significant though was that in the same year Arnold Weinstock joined GEC as chief executive when the company took over Sobell, his father-in-law's TV manufacturing firm and he then set about a rapid program of mergers and rationalizations, which led to GEC eventually becoming the biggest industrial conglomerate in the United Kingdom.

In the 1960s, there was considerable dissatisfaction between the supply industry and the GPO, and it was clear to the suppliers that a better path was needed to reequip the U.K. telephone system than the big bang expected of immediate adoption of electronic exchanges. To this end, Plessey Liverpool, as it then was, had started a privately supported development of Crossbar technology because it felt that it was going to be necessary to have interim solutions to equip exchanges as the overall size of the telecoms market was expanded. The first sale of Crossbar was made in 1968 to a rather reluctant GPO, with Plessey being required to license its system to GEC.

Meanwhile, Arnold Weinstock at GEC had begun the process that was to make GEC the dominant electronics company that it became and a hotly contested takeover by GEC of AEI ensued in 1966. This included the GEC taking over the Woolwich telecoms factory, which was soon scheduled for closure with 5,000 redundancies. Although strongly resisted, Weinstock had industrial logic on his side and the factory was poorly placed both technologically and financially to argue a case against closure. The Woolwich factory was the earliest surviving telecoms factory in the United Kingdom. It had eventually become part of the AEI conglomerate. In line with what was accepted as the poor governance applied by AEI (which led to their takeover by GEC), the factory had been largely left alone and underinvested. The unexpected loss of £3 million in the AEI telecoms results had led to the takeover battle swinging in favor of GEC. At the time of closure, it was claimed that the output per employee was around £500 per annum, a spectacularly low figure even by the standards of the day [37]. The closure happened in 1967 and meant that there were now only three suppliers to the GPO Bulk Supply Agreement, Plessey, GEC, and STC, but amazingly the share of the items supplied was kept to the old five-company model, so Plessey received 40%, GEC received 40%, and STC received 20%.

In 1968, GEC carried out another merger, this time the agreed takeover of the English Electric company. The main impetus for this was to carry through the rationalization of the electrical power industry, which duly took place with GEC controlling three of the major players namely AEI, English Electric, and GEC itself. However, GEC discovered that the jewel in the crown was the Marconi Company (another company that was named after a foreign entrepreneur but had long since ceased to have any links outside the United Kingdom), which was owned by English Electric. This company had excellent technology and could put that into the telecoms development although Marconi, mainly based at Chelmsford, was never anything other than a minor supplier into the U.K. telecoms industry.

Rather perversely, the GPO also used TMC (the Telephone Manufacturing Company, sometimes referred to as Temco) as a minor supplier of telecoms switching equipment. TMC was acquired by Pye (an early innovator of U.K. radios) in 1960, which, in turn, was acquired by the Dutch conglomerate Philips in 1967. The company had set up two factories making Strowger equipment in Malmesbury and Greenwich. While this might have provided an attractive alternative supplier to the industry, there seems to have been little action to bring in the technology available within Philips.

The issue of the poor productivity of the U.K. telecoms equipment industry was well known. In his 1976 STC communications lecture, Sir William Ryland, then chairman of the Post Office stated:

> Western Electric made last year 4 and a half million lines of electronic switching and they did it in five factories. We now have in this country a capacity to produce 1 million lines of electronic switching and it is just dispersed over 9 towns—not factories but towns.
>
> So very broadly we are now equipped to do about a quarter of what they do but we need twice the number of locations in which to do it. Hardly the way to keep a unit costs in check.

Sir William was looking at a supply industry in which the United States were well ahead of the United Kingdom in the introduction of reed relay switches. His comments were well informed because in September 1976 there was a visit of Post Office engineers to their counterparts in Western Electric. Among the data (held in the BT Archives) are reports that showed that the U.S. telecoms industry was more productive than that in the United Kingdom and that its productivity would be dramatically improved by technology change. The visitors were given detail access to productivity figures, which indicated that, for electromechanical switches, U.S. factories had achieved about a 20% lower cost per line than the then-current U.K. cost of Strowger (about £50 per line). The Western Electric Oklahoma City plant had largely converted to the ESS reed switching exchange and had seen productivity (measured in terms of output of lines per employee) treble between 1970 and 1976 with an output of over 2 million lines per annum exceeding the capacity of the entire U.K. supply industry. While the costs of electronic switches were initially much higher than electromechanical switches, there was a widespread belief that the declining costs of components would drive these costs down below those of existing electromechanical switches [39]. The cost of transistors had been at least halving every 10 years [40].

Because of the late introduction of replacements to Strowger switching, the U.K. industry had to continue to rely on it to help address a chronic undersupply situation that existed in the 1970s. The supply shortfall in the period 1978 to 1980 was partly a consequence of treasury-driven cutbacks in 1975 when the U.K. industry was supplying around 1.5 million lines per annum split mostly equally three ways between Strowger, Crossbar, and TXE4A (reed relay) exchanges, according to the BT Archives. The accelerating introduction of TXE4A exchanges in the late 1970s meant that, by the time of privatization in 1984, there was only a small residual waiting list [41].

With the expected widespread adoption of electronic switching and the increasing use of electronics in data transmission the electronic supply industry had to face the issue that they had many electromechanical factories that were obsolete. In practice, it was easier to shut down these factories and buildup capabilities in electronics in other locations.

In 1975, there were approximately 50,000 people employed in making telephone equipment in the United Kingdom. Of the factories that existed in the United Kingdom making electromechanical switches in the 1960s, most were satellite factories that developed using funding by government investment incentives in deprived areas where traditional industries (such as textiles and coal) were declining such as Northern Ireland, mid-Scotland, the north of England, and South Wales. In Appendix B, the main factories of the U.K. telecoms industry in 1975 are listed showing 20 satellite factories out of the total 35 sites [29]. These satellite factories accounted for about half the U.K. workforce, and almost all of these were shut down in the 1980s. Thus, while the output of the telecoms supply industry increased from the 1970s to the 1990s (in terms of lines supplied), its employment fell dramatically to less than one-half the workforce in 1975.

The need for the telecoms industry to develop capabilities for digital switching also meant they had to dramatically change the entire configuration of their operations. The electromechanical exchanges predominantly used in-house produced parts with a relatively low external component cost. However, digital electronics production was quite the opposite with a large array of external components required. This included printed circuit boards (PCBs), power supplies, various passive components, but, most of all, semiconductors. The industry made various steps to improve the supply situation with all these components including starting in-house manufacturing. However, the biggest challenge was to up their game in the supply of semiconductors. While a detailed analysis is outside the scope of this work, the U.K. industry did not respond well particularly to the adoption of the silicon planar epitaxial process that enabled large-scale integration of semiconductors to be started in the 1980s onwards. Although the main suppliers such as GEC, Plessey, and STC all had substantial semiconductor capabilities, these were inadequate compared with the scope of the challenges that they faced. Technology was often licensed in from overseas suppliers, but the choice of technology was sometimes more to do with the price of the license than the quality of the technology [41]. Despite having considerable financial resources (at least in the case of GEC), few attempts were made to buy out a major semiconductor house, which would have dramatically improved the capability of the industry.

From an employment/governmental point of view, the industry was subject to a dramatic decline since all the electromechanical production would be rapidly terminated and the corresponding growth in electronics would, to a fair extent, come from externally supplied components, many of which were imported. This would have been the case even if the industry could develop successfully its own digital switching capability.

The privatization of BT had a significant effect on the evolution of the U.K. supply industry. By the time of the privatization, System X was still in development. While initial deliveries were made in 1981, no significant installations of System X were achieved until 1985, and, at the same time as this, BT made its first purchase of a competing system that it called System Y. This was the Ericsson AXE system that was being offered in the United Kingdom by a joint company, Thorn-Ericsson (Thorn being a long-established U.K. electronics group with little telecoms presence) [35]. The excitement of the privatization of BT and its removal from political pressure meant that there was little that could be done politically to stop this process. Indeed, BT was widely encouraged to adopt an aggressive approach to it suppliers to procure the best-of-breed systems at the lowest cost.

The U.K. industry was in a poor position to meet this demand and, with the BT purchase of a competing system, the effective revenue available to the industry would be dramatically reduced by the late 1980s. The number of lines bought from the traditional three main suppliers, GEC, Plessey, and STC, was eroded by purchases of the new AXE system from Ericsson of Sweden that had courted BT for many years and established a factory in Scunthorpe, Yorkshire, to supply the equipment. The effect of technology and competition was driving down the unit prices and increasing the bought-in content of their products. While demand for fixed-line connection in the United Kingdom had steadily increased from 1950, the market was very well saturated by 2000 (Figure 2.6), so all that remained for the supply industry was to replace obsolescent mechanical switching exchanges. The industry thus had to dramatically reduce its workforce. The traditional electromechanical product factories had little in common with the needs of the new electronic systems. Thus, it was inevitable that many of the satellite factories set up in the 1960s would be closed (see Appendix B). Sites shut included: GEC; Middlesbrough, Treforest, Aycliffe, Hartlepool, Kirkcaldy, West Chirton, Plessey; South Shields, Ballynahinch, Kirkby, Wigan, Chorley, Huyton, Sunderland STC; and Monkstown, Larne, Enniskillen, East Kilbride, Treforest, Benfleet.

I calculate that the aggregate employment in the telecoms supply industry (excluding BT) fell from around 50,000 to under 20,000 in the period from 1975 to 1990. Ironically, many of the job losses were in the traditional industrial areas that were at the same time suffering from the losses of jobs in coal-mining, steel, and other manufacturing resulting from the "Thatcherization" of the British economy (see Appendix B).

However, with relatively poor export performance of the industry, a major factor was the longer operating life of electronic exchanges, such as System X, compared with its preceding systems. I estimate that the average system life of an exchange in the U.K. network was:

- Strowger: 15 years;
- Crossbar: 13 years;
- TXE4A: 10 years;
- System X: 20 years or more.

These life spans (calculated on stated installation and replacement rates) go some way to support the (then) GPO's reluctance to install interim solutions before a fully functioning electronic exchange was available. However, given the relatively static demand for phone lines, it is also clear that an industry that was at its peak supplying almost 4 million lines per year had to adjust to a situation where the steady state requirement would fall to below 1 million lines. In practice, System X represents the last of the "traditional switching exchanges." Telecoms networks are now almost exclusively designed around packet switching using TCP/IP based transmission. BT's announcement that it will have a full fiber network by 2025 [42] means that their System X exchanges (which performed well in service) are planned to be switched off by then.

The long overdue rationalization of the industry became inevitable. STC had already been given a broad hint about its future when it was carved out of the System X Consortium in 1976. It had been granted a major role in the supply of the interim exchange TXE4A with the delays in System X; orders for

TXE4A were strong in the period from1979 to 1985 averaging over 1 million lines per year. This boosted STC's financial performance and enabled them to mount a takeover of the United Kingdom's biggest computer company, International Computers Limited (ICL). The rationale behind this move was the intention to provide an integrated solution with the convergence of communications and computing. In practice, the two companies were badly matched. No great synergy was ever achieved. STC eventually was acquired by Nortel, a Canadian telecoms competitor that, like STC, had its roots as a subsidiary of the U.S. AT&T monolith. This takeover was made in 1990 in conjunction with the Japanese Fujitsu company acquiring most of the remaining ICL business (see Chapter 5). Ironically, Nortel itself fell a victim to overstretching itself in the post-dotcom crash and eventually filed for bankruptcy in 2009.

Of even more significance was the takeover of Plessey by GEC. Although Plessey had built up substantial cash resources in the late 1980s partly as a result of its work on the System X project, its resources were dwarfed by those available to GEC (at the time the strongest industrial company in the United Kingdom) with a fabled £1 billion cash mountain [43]. It was clear that with the falling demand for their telecoms equipment these resources would be eventually depleted. In 1985, GEC proposed the takeover of Plessey, but this was blocked by the government because of the monopoly that it would have given GEC in the supply of military electronics equipment, both companies being major suppliers in this area [44]. As an interim arrangement, the telecoms interests of GEC and Plessey had been put into GPT (GEC Plessey Telecommunications) in 1987 equally owned by GEC and Plessey in an uneasy alliance. However, this action to recover the situation with telecoms equipment was going to be too little too late.

However, GEC was not going to go away and eventually the full takeover of Plessey was consummated in 1989 by the maneuver of bringing in the German conglomerate, Siemens, as a partner in the acquisition and carving Plessey up in a way that made the government reasonably happy. As part of the 1989 deal, Siemens took a 50% share of the merged GPT telecoms businesses. The relationship between GEC and Siemens, although reasonably good, was not strong enough to really affect the business, with there being little willingness of Siemens to inject its technology into the U.K. market.

In 1999, as part of a dash for growth, with its new CEO, George Simpson, GEC, bought out the Siemens interest in GPT and rebranded itself as Marconi to focus exclusively on telecoms. The key actions involved are described in more detail in Section 5.4. However, disastrous acquisitions provided disappointing performance and the share price of Marconi collapsed. The company had virtually ceased to be a serious player by 2002, and the remaining pieces of the telecoms business were bought out by competition in the following few years [45].

Thus, within 17 years of the privatization of BT, there was no Tier 1 supplier of telecoms equipment left in the United Kingdom, all supply being through subsidiaries of other multinational organizations. BT became increasingly enthusiastic about buying equipment on the world's market from suppliers such as Ericsson and Huawei, the upcoming Chinese supplier. The relationship with Huawei came to an abrupt halt with the end of the honeymoon period of relations between Britain and China overseen by the U.K. government of David Cameron (2010 to 2017). This came to an end with the crackdown in Hong Kong and the subsequent concern about letting China into what was belatedly seen as a key strategic asset [46].

References

[1] Valdar, A., "Circuit Switching Evolution to 2012," *Journal of the Institute of Telecommunications Professionals*, 2012.

[2] "The Growth of Telecommunications Services in the United Kingdom," *The Post Office Electrical Engineers' Journal*, 1956, pp. 161–165.

[3] Carter Committee, *Report of the Post Office Review Committee, Cmnd. 6850*, Section 3.5–3.8, London: HMSO, 1977.

[4] Harper, J., *Monopoly and Competition in British Communications*, London: Pinter, 1997, pp. 95–100.

[5] Young, P., *Power of Speech: A History of Standard Telephones and Cables, 1883–1983*, London: Allen & Unwin, 1983, p. 76.

[6] Young, P., *Power of Speech: A History of Standard Telephones and Cables, 1883–1983*, London: Allen & Unwin, 1983, p. 148.

[7] Valdar, A., *Understanding Telecommunications Networks*, London: IET, 2006, p. 56.

[8] Harper, J., *Monopoly and Competition in British Communications*, London: Pinter, 1997, pp. 199–201.

[9] Young, P., *Power of Speech: A History of Standard Telephones and Cables, 1883–1983*, London: Allen & Unwin, 1983, p. 176.

[10] "F.C.C. Authorizes Fiber Optic Project," *New York Times*, May 25, 1984.

[11] Horak, R., "A Brief History of the Fax," *Telecom Reseller*, March 2, 2010.

[12] Harper, J., *Monopoly and Competition in British Communications*, London: Pinter, 1997, p. 5.

[13] Young, P., *Power of Speech: A History of Standard Telephones and Cables, 1883–1983*, London: Allen & Unwin, 1983, p. 63.

[14] *Internal Reports on Exchange Policy*, file ref. TCB 712/3/7, London: GPO, 1976.

[15] *Internal Documents on Exchange Strategy*, file ref. TCC 333, London: GPO, 1972.

[16]   Harper, J., *Monopoly and Competition in British Communications,* London: Pinter, 1997, p. 49.

[17]   Thatcher, M., *The Politics of Telecommunications: National Institutions, Convergences, and Change in Britain and France,* Oxford, U.K.: Oxford University Press, 2000.

[18]   Institute for Government, *Privatisation of British Telecom (1984),* 2018.

[19]   Carter Committee, *Report of the Post Office Review Committee, Cmnd 6850,* London: HMSO, 1977.

[20]   "Hotline to Cable & Wireless," *Economist,* 1981, pp. 109–110.

[21]   "Virgin Media Bought for £15M," *Guardian,* February 6, 2013.

[22]   "Lechleider Dies at 82," *New York Times,* May 3, 2015.

[23]   Woods, B., "Challengers Fall Behind BT in Superfast Broadband Race," *The Telegraph,* November 1, 2021.

[24]   Thatcher, M., *The Politics of Telecommunications: National Institutions, Convergences, and Change in Britain and France,* Ch. 7, Oxford, U.K.: Oxford University Press, 2000.

[25]   Ward, K. E., "The Inland Network," *Post Office Engineering Journal,* 1981, p. 165.

[26]   Ritchie, B., *Into the Sunrise: a History of Plessey 1917–1987,* James & James, 1989.

[27]   Central Policy Review Staff (CPRS), *Telecommunications Manufacturing Industry,* Government Cabinet Office, 1978.

[28]   Aris, S., *Arnold Weinstock and the Making of GEC,* London: Aurum, 1990.

[29]   TUC, *Statement on the Future of the Telecommunications Industry,* London: Trades Union Congress, 1977.

[30]   Healey, D., *Ministerial Group on the Telecommunications Manufacturing Industry -Minutes,* London: Cabinet Office, October 31, 1978.

[31]   Harper, J., *Monopoly and Competition in British Communications,* London: Pinter, 1997, p. 124.

[32]   Knapp, H., *System X, Report to Sir Kenneth Berrill, Head of Cabinet Office,* Ref. CAB184_375_05, Central Policy Review Staff, UK.gov, 1978.

[33]   Harper, J., *Monopoly and Competition in British Communications,* London: Pinter, 1997, p. 123.

[34]   Monopolies and Mergers Commission, *GEC & Plessey: A Report on the Proposed Merger, Cmnd 9867,* London: HMSO, 1986, p. 32.

[35]   "Rules of the New Telecoms Game Change Every Day," *Economist,* July 27, 1985.

[36]   Johnstone, B., "£2,500M System X Project Starts," *Times,* September 12, 1980.

[37]   Aris, S., *Arnold Weinstock and the Making of GEC,* London: Aurum, 1990, p. 72.

[38]   "Columbian Order for Plessey's System X," *Times,* March 10, 1987.

[39]   Jang, S. -L., and J. R. Norsworthy, "Productivity Growth and Technological Change in the United States Telecommunications Equipment Manufacturing Industries," in M. Crew,

Competition and the Regulation of Utilities: Topics in Regulatory Economics and Policy Series, Vol. 7, Boston, MA: Springer, 1991.

[40]   Malerba, F., *The Semiconductor Business: The Economics of Rapid Growth and Decline*, Madison, WI: University of Wisconsin Press, 1985, p. 26.

[41]   Malerba, F., *The Semiconductor Business: The Economics of Rapid Growth and Decline*, Madison, WI: University of Wisconsin Press, 1985, pp. 63–69.

[42]   Woods, B., "Challengers Fall Behind BT in Superfast Broadband Race," *Telegraph*, November 1, 2021.

[43]   Aris, S., *Arnold Weinstock and the Making of GEC*, London: Aurum, 1990, p. 168.

[44]   Aris, S., *Arnold Weinstock and the Making of GEC*, London: Aurum, 1990, pp. 190–196.

[45]   O'Connel, D., "Marconi Wipeout Puts Shareholders in Gloom," *Sunday Times*, Business Section, September 1, 2002.

[46]   Flides, N., "Huawei Fall Sees Nokia Become BT's Top Supplier," *Financial Times*, September 30, 2020.

3

The Mobile Revolution

Mobile telephony made a huge change to society in the last 20 years of the twentieth century. By 2000, just about everybody in the developed world possessed a mobile phone and even developing countries were rapidly expanding the ownership of phones. The 1980s saw the start of this major change in telecoms, which bypassed much of the U.K. telecoms supply industry. Mobile radio communications had been widely used in World War II, and, after the war, emergency services and some private users introduced simple wireless networks. The increasing wealth and mobility of populations made the potential demand for mobile phones clear from the 1940s. However, it required the invention of the transistor in 1947 (with the subsequent development of the microprocessor) and the resulting dramatic reduction in the size and cost of radio communications to make mass mobile telephony a feasible proposition. Few people at the time understood the dramatic changes in society that would follow on from this innovation as it joined with the development of the internet to produce today's mobile information society.

In this chapter, we explore the history and the effect on the industry of the development of mobile telephony. First, we look at how the technology was developed and introduced. Second, we look at how regulatory standards were established and applied to various countries, contrasting the experience of the United Kingdom with other European countries, the United States, and Japan. With the development of international standards, suppliers were able to address a global market. Third, we look at how service providers were established and evolved and how they interfaced with the telecoms supply industry. In conclusion, we look at how the U.K. telecoms industry performed both as a supplier of mobile phone services to consumers but crucially as suppliers of equipment to the newly established networks.

3.1 Mobile Technology

Commercial radio had been available since early in the twentieth century and mobile communications by radio was also explored, initially with shipping and then a trial of a mobile radio system on the German railway system in 1918. Radio communications were an important factor in World War II, particularly for mobile forces. Naturally, this concept was explored in civilian applications postwar to address communications issues where fixed-line telephony was not available. The growing availability of transistors in the 1950s enhanced this trend, but, in most cases, radio communications were restricted to point-to-point arrangements. In the 1940s, various mobile communications networks were developed, often for vehicles. The MTA system was introduced in Sweden in 1956, while AT&T in the United States launched the Improved Mobile Telephone Service (IMTS) in 1964. Although an improvement from its earlier Mobile Telephony Service, it still suffered from limited capacity and only achieved 40,000 subscribers. With their limited capacity and high cost to the subscriber, uptake on these networks was low.

Another early development of mobile communications was the development of pagers. These were often just a simple device that would be carried by individuals and produce a buzz or vibration to notify them to contact a message center. They were widely used in organizations from the 1960s onwards and remained in use in specialized applications (such as hospitals and emergency services) up until the 2020s [1]. The pager was patented by U.S. inventor Alfred Gross in 1949; early versions used independent radio transmission covering a relatively small area. In 1962, Bell launched the Bellboy pager in the United States, which had a wide coverage with a central messaging center for people to call if paged. Later, more sophisticated pagers were manufactured by companies such as Motorola. These could feature two-way text messaging. However, the increasingly sophisticated functionality of mobile telephony led to decline in usage of separate pagers, and most manufacturers stopped making them around 2000.

In a similar way, the development of cordless handsets for telephones was extended in some countries into the provision of a handset that would link up to different base stations, thus providing some of the facilities later seen in cellular mobile phones. These handsets often used the CT2 protocol that provided for duplex operation with a digital FDMA system. In several countries, including the United Kingdom and France, a national network was offered, often referred to as Telepoint, usually linked to commercially provided base stations. An example of this is the Rabbit network operating in the United Kingdom from 1992 to 1993 by Hutchison, which aimed to have 12,000 base stations, often sited at the Little Chef café chain [2]. However, the growing use of cellular phones with their wider area coverage and much superior handover between base stations limited the market for Telepoint, and most of these services had ended by the end of the 1990s. Cordless handsets for landlines continue to be used to this day, often using the Digital Enhanced Cordless Telecommunications (DECT) protocol that superseded CT2.

3.1.1 The Cell Phone

The basic issue of mobile radio communications is how to provide wide coverage and sufficient channels to accommodate many users at a reasonable cost. The adoption of a cellular system architecture resolved these issues for mass mobile telephony. The concept of a cellular telephone switching plan and the crucial handover procedure between cells was largely pioneered by AT&T in the United States and was described by Amos Joel Jr. [3]. Ironically, AT&T did not invest significantly in commercializing the idea since their market research showed only a small potential market. The basic concept of cellular radio can be shown in Figure 3.1 as a 7-cell pattern concept. Each cell color represents a small geographic area (often a few kilometers or less across) each with a specific set of channels within an allocated transmission frequency band. These bands would carry many channels, dynamically allocated to individual users [4].

Because signal power declines as the square of distance from the transmitter, cells coded with the same color above (hence the same frequency) do not interfere with each other. Thus, a large number of users can be accommodated without the use of a vast number of radio channels. As a user moves from one cell to an adjacent cell, there needs to be some handover arrangement whereby the phone hands over the channel that it was using in the previous cell and is allocated a new channel in its current cell. This concept was introduced in a paper presented to the U.S. Federal Communications Commission (FCC) in 1971.

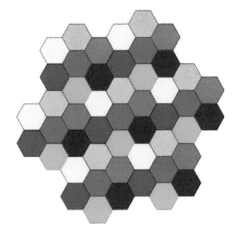

Figure 3.1 Seven-cell mobile arrangement. (Source: Public domain; copied from Valdar [4].)

This system was further developed and described by AT&T's Fluhr and Nussbaum in 1973 [5]. The concept of cellular architecture enabled several subscribers to communicate through a common cell (or central transmitter) to each other. A logical extension of this is the development of a network of adjacent cells where a subscriber could log into the appropriate cell for their location and communicate with any other subscriber who was also logged on to the network. This concept, shown in Figure 3.2, evolved into the basic architecture of virtually all mobile telephony.

It became apparent with the increasing sophistication of electronic components that were becoming available in the 1970s that hand-portable devices were feasible. Various companies had developed the technology and, in the United States, Figure 3.3 shows what is widely regarded as the first portable mobile in 1973 made by Motorola and weighing in at 1.1 kg [6].

The first cellular network for mobile phones opened in Tokyo by NTT in 1979 [7]. Once the basic concept of mobile telephony had been demonstrated, the way was open to spread the technology across the world. The Nordic countries developed a unified protocol (Nordic Mobile Telephone (NMT)), which was rolled out in their countries in 1981. The United States introduced its first cellular mobile network in 1983 and, by the mid-1980s, many other countries had introduced a network (including the United Kingdom). The clear market potential of everyone having a mobile phone ensured private capital was available to fund the necessary development of infrastructure and equipment. The key ingredients would be:

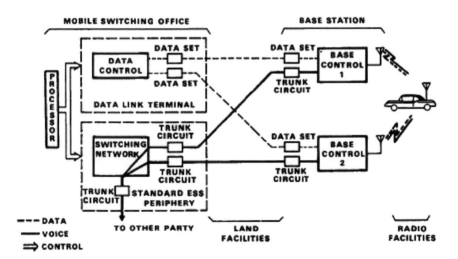

Figure 3.2 This schematic is taken from the original paper published by Fluhr and Nussbaum of AT&T [5].

Figure 3.3 Martin Cooper photographed in 2007 with his 1973 handheld Motorola mobile phone prototype. (Source: Rico Shen CC-SA 3.0.)

- *Handset supply:* This is a high-volume consumer application that was addressed both by existing suppliers in related applications and by new start-ups. Early suppliers were:
  - Established suppliers of handheld radios and similar equipment. This would include Motorola (United States), Hitachi (Japan), and Sony (Japan).
  - *Telephone equipment suppliers:* This would include Ericsson (Sweden) and Siemens (Germany).
  - *Start-ups:* The most notable is Nokia (Finland).
- *Network infrastructure:* Although the transmission masts were a new requirement, the transmission between masts and the associated switching was closely related to the existing fixed telephony network. Thus, in many countries, existing PTTs assumed a monopoly position in the early supply of mobile telephony. This would be the case in Japan and many European countries.
- *Service providers:* Again, the natural service suppliers would be national PTTs since they had both the links to national regulatory providers, the fixed-line communications infrastructure, and the existing customer

base. What is remarkable is that, due to political and business pressures, this did not always prove to be the case.

• *Regulatory authority:* In virtually all countries, the availability of bandwidth for the transmission of radio signals is closely controlled by some form of national regulatory body. Sectors of bandwidth are usually allocated to certain applications, such as radio and TV broadcasting or military or commercial use for specified applications. While some sectors of bandwidth had been left open for amateur communications (and other personal media such as CB radios), these were relatively small. For cellular mobile telephony to be taken up by many users, a significant bandwidth would have to be allocated to this purpose by the regulator and licensed to service providers. Fortunately, the rollout of mobile telephony largely coincided with the freeing up of some TV bands as TV in the United Kingdom and much of Europe was transitioning into higher-definition color pictures that needed higher bandwidth, which, in turn, required the signals to operate at a higher frequency.

In the 1980s and onwards, all these pieces came into place and the rollout of mobile telephony began. The Organisation for Economic Cooperation and Development (OECD) statistics in Figure 3.4 show the penetration of mobile phones in key countries.

As can be seen, in 1988, numbers just about register as a few percent in the Nordic countries, but, by 1998, most developed countries had over 20% penetration. By 2008, mobile phones had become ubiquitous with many countries having more mobile phones than population.

Most of the early development of mobile communications was based on voice traffic, but there was also a smaller demand for data communications for specialized applications such as security. In several countries, separate networks were established for paging and other data applications. This used technology that was very like mobile phones but was designed for carrying short text messages. While initially messages were simple notifications to an individual device, the technology was extended to cover applications such a burglar alarms. In many countries, such as the United Kingdom, these applications were achieved by rolling out entirely separate networks and there were 800,000 pagers in use in the United Kingdom in 1994 [8]. While the later development of data transmission via mobile phones would restrict the market for these networks, some of them still exist.

Early development of mobile telephony was usually based on regarding each country as a closed area that would be free to develop its own products and standards based on its own circumstances. However, conflicting pressures

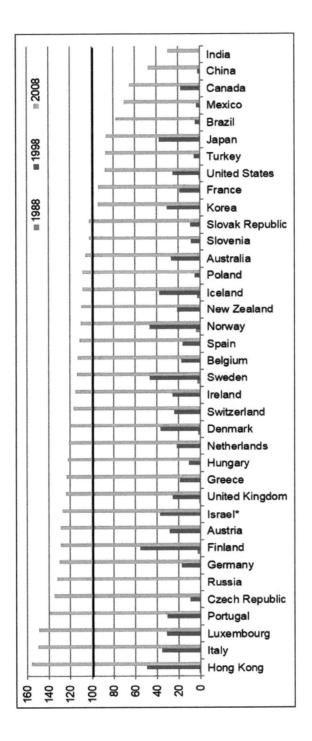

Figure 3.4 Number of mobile phones per 100 population, in 1988, 1998, and 2008 by country. (Data from UNESCO.)

would eventually globalize the market to the (almost) common standards that enable worldwide roaming of mobile phones. Factors included:

1. *The need to accommodate countries with high levels of interaction.* A notable example of this is Scandinavia where close collaboration on mobile telephony rollout was necessary from the beginning because of both the small size of each market and the high level of movement within the overall area.

2. *The economies of scale in production of handsets and other equipment.* Consumers would be highly price-sensitive in the purchase of handsets, which are very complex pieces of equipment. International suppliers could offer more attractive prices as volumes increased. This could be greatly helped if their handsets worked in several countries (and hence met certain transmission standards).

3. *Common bandwidth.* Much of the early spectrum of mobile phones came from spectrum vacated by TV transmission. The need to avoid interference across national borders and the commercial pressures mentioned above, which also applied to the rollout of TVs from the 1950s, meant that many countries would use common frequency ranges.

Thus, the mobile phone market rapidly became a world market with alignment in specifications in Europe and the Far East. Because of the geography, size, and economic strength of the United States (and Canada), the alignment was not so strong there and this situation persisted until recently. However, the increasing sophistication of phone handsets plus the availability of networks in North America operating to international standards compatible with GSM (albeit at different frequencies) has prevented this from being a major issue.

3.2 Phone Generations

Regulators often have a bad reputation, being seen as bureaucratic necessities often restricting the natural dynamism of the free market. It can be argued that, for mobile telephony at least, the opposite is the case. We shall see how the tightly regulated markets in Europe and Asia spawned much of the early successes in mobile networks while the more loosely regulated United States was left behind (at least until the social media revolution arrived after 2000).

In this section, I shall expand on how the specifications for mobile phones evolved and the consequent rollout of mobile phone networks throughout the world. The various protocols and specifications generated a veritable alphabet soup of acronyms. Most of these are described briefly in the text but are expanded upon further in the Acronyms and Abbreviations. The specification of the various phone generations is complex; in this section, I summarize some of the key features, but for a more detailed description, read Chapter 9 of [4]. These different generations can be summarized in Table 3.1 (giving the key data for each protocol in both the United Kingdom and United States).

3.2.1 First Generation (1G): Analog Phones

The initial design of mobile phones was based on analog radio transmission practices and was not heavily standardized with regard to intercountry working. Nor was it very sophisticated, so it could be hacked relatively easily. However, it laid the basic foundations for the creation of the new market and established the feasibility of a large proportion of the population owning and using mobile phones.

At this early stage, several noncompatible standards emerged. Nordic countries were early in introducing a common standard, NMT, which enabled the world's first cellular roaming across borders. In the United States, phones used the Advanced Mobile Phone System (AMPS) standard, which was introduced in 1983. The United States mainly used a frequency band around 800 MHz compared with 900 MHz used in most of Europe. The United Kingdom adopted its own standard, a derivative of the U.S. AMPS standard called Total Access Communications System (TACS) in 1983. Other European countries, including Germany, France, Italy, and Spain, also generated their own standards with little compatibility between countries. As a pioneer, Japan deployed several

Table 3.1
Main Phone Generations in the United Kingdom and the United States

Generation	Known As	Introduced (U.K./U.S.)	Transmission Frequency (MHz)	Data Rate (kbps)
1G	Analog	(1985)/1983	(900)/800	NA
2G	(GSM)	(1992)	(900/1,800)	(50)
2G	CDMA	1995	800/1,800/1,900/2,500	—
2½G	(GPRS)	(2000)	(900/1,800)	(300)
3G	(WCDMA)	(2002)	(1,800)	(384)
3G	EV-DO	2002	1,800	300
3½G	HSPA	2005	1,800	1,800
4G	LTE/WiMAX	(2012)/2010	700/800/(900/1,800)/1,900 (2,100)	4,000
5G	Low-band 5G	2019	900	30,000
5G	Midband 5G	2019	1,700–4,700	100,000

Shaded cells represent common data, parentheses represent U.K. data, and U.S. data is without parentheses.

standards before introducing a variant of the U.S. standard (Japanese Total Access Communications System (JTACS)).

While the switching information between cells was digital, in most cases, the phone signal itself was analog narrowband-frequency modulation (FM) modulated onto a carrier and hence it could be relatively easily hacked. The first viable network established was in Japan in 1979, but it was particularly notable that the Scandinavian countries managed to agree with a standard to their networks and achieve an early lead in terms of phones per head of population.

In Europe, France and Germany tried to champion a common standard, but internal rivalry led to this breaking down. Notably, the German system pioneered the use of the Subscriber Identifier Module (SIM) card in its not-very-successful N specification. SIM cards were to become a standard feature of mobile phones. Thus, throughout Europe, early users of mobile phones would have little chance of them working beyond their home country; in the United States roaming between carriers was difficult because of the commercial policies of the carriers.

As one would expect in a newly developing market, local suppliers held a strong position, particularly where the specifications were unique to the territory as was the case in the United States, France, Germany, and Japan. However, in some cases, such as the United Kingdom, the lack of local availability of key items (handsets and transmission equipment) meant that international suppliers such as Nokia, Ericsson, and Motorola were able to dominate the supply situation.

3.2.2   Second Generation (2G): Europe Wins the Battle to Create a Global Digital Standard

By the mid-1980s, it was clear that there was a major new market emerging and the existing transmission standards were inadequate. Market demand had been demonstrated by the analog (1G) networks, but there were significant issues with the lack of international compatibility, poor use of bandwidth, the absence of any digital services, and ease of hacking. Many people believed that a fully digital network was required. With this would come the introduction of the now ubiquitous Short Message Service (SMS) text messaging service.

In Europe, low-key discussions had taken place from 1982 under the umbrella of an organization CEPT (a French acronym for an organization of European PTTs). A growing interest in standardization within Europe would lead to the single market being adopted by what became the European Union (EU) in 1992. The group discussions began to gather pace under the title Groupe Speciale Mobile (GSM) and the United Kingdom joined the discussions in 1984 and signed up to full involvement in the GSM process in 1986. With the expansion of the EU, there was a lot of interest in how standardization within the biggest market block in the world could benefit the consumer. The GSM development fitted this need ideally and thus received an enormous political boost when in 1987 the council of European Communities issued its recommendation 87/371/ EEC on the introduction of cellular mobile communications within the community. Given this impetus, in an extraordinary example of cross-border collaboration the group prospered and eventually agreed to a new specification for mobile phones based on narrowband time division multiple access (TDMA), which was adopted throughout Europe and eventually throughout the world. By 2010, GSM had 4 billion subscribers using this protocol.

A view of the sheer complexity of the technical discussions and the political issues that had to be resolved is given in the self-published account by a U.K. representative on the GSM, Stephen Temple [9]. The specification was based on technical work between France and Germany, but this had also received a significant input of British pragmatism. As well as helping to resolve which technical design to follow, the United Kingdom provided a perspective of having two private network operators in the country compared with the state monopoly operated by France and Germany at the time. It uses digital transmission of narrowband TDMA and a frequency band around 900 MHz with a later addition of 1,800 MHz. The important thing was that the specification was fit for purpose and would be widely adopted. For the first time, the roaming capability of GSM meant that mobile phone users could use the same phone throughout the world (see Figure 3.5). Even in the United States, GSM networks would become available, although it was less dominant there.

The first 2G, GSM network was introduced in Finland in 1991. In the United Kingdom, the first network was introduced in 1992.

Although envisaged as a standard for countries within the European Economic Community (EEC), the availability of a proven standard for digital mobile phones together with an established supply base was attractive to many other countries. Early adopters of GSM for mobile telephony included Australia and Thailand and, by 2005, GSM networks accounted for more than 75% of the worldwide cellular network market, serving 1.5 billion subscribers [10].

The United States did not engage in the development of the GSM standard and produced its own IS95 protocol based on code division multiple access (CDMA), which was rolled out in the United States in 1995 and in some other countries, notably Canada and South Korea, and some South American and Eastern European countries. However, even the United States had to accept the global strength of GSM and networks compatible with GSM were introduced into the United States (albeit as a niche market). The increasing interest in using mobile data was becoming a major issue. GSM was relatively poor in this regard achieving data rates up to 50 kbps. Increasing demand for data transmission led to the adoption of the GPRS and EDGE protocols to enable GSM phones to achieve more acceptable data rates (around 300 kbps).

Figure 3.5 The Nokia 1011 widely accepted as the world's first commercial GSM mobile phone. The Nokia 1010 family of phones (despite have long gone out of production) is still probably the world's most widely sold phone with approaching 500 million sold. (Source: Nokia Corporation CC BY-SA 4.0.)

The European-wide acceptance of the GSM specification meant that European-centered suppliers, such as Nokia and Ericsson, were in a strong position to supply the market in Europe and beyond. In the United States, local suppliers such as Motorola provided a lot of equipment and handsets, but Asian suppliers began to seriously penetrate this key market.

3.2.3 Third Generation (3G): Mobile Data Becomes Important

The huge success of 2G mobile phones led to inevitable problems. The demand for phones made securing more bandwidth vital. While the main initial driver for demand was voice calling, the demand for digital services was becoming more significant. Initially, 2G networks were largely designed for voice traffic and although digitized using TDMA or CDMA techniques voice connections used conventional telephony to interconnect. The initial 2G phones had a low data-rate for data applications (around 50 kbps), later specification enhancements using GPRS (2½G), and EDGE (2¾G) increased this to around 300 kbps, but this only highlighted the need to redesign the phone protocols for a mix of voice and data traffic. The 3G specifications were thus conceived of using packet switching for both voice and data traffic.

The market need, in turn, generated a plethora of protocols, many of which arose from the work of 2½G and 2¾G implementations. In particular, some major carriers in the United States persisted with its CDMA-derived specifications, using a derivative termed CDMA2000-EV-DO. A competing specification WCDMA (a successor to GSM) was used in Japan and Europe. Japan was first to implement a 3G network in 2001 while the United States and parts of Europe including the United Kingdom first launched networks in 2002. Further enhancements to the specifications were required to achieve the data rates necessary for video linkages (1 to 10 Mbps) and the HSPA protocol was introduced. By 2007, there were reportedly 295 million subscribers on 3G networks, with WCDMA having around 66% of the market and balance being the EV-DO standard.

In this dash for bandwidth and capacity to support 3G rollout many governments, led by the United Kingdom, found a novel way to enhance their taxation revenues by auctioning bandwidth to existing and incoming service providers. This was facilitated in the United Kingdom by the 1998 Wireless and Telegraphy Act and a complex and keenly fought auction was arranged for what was to become five licensed operators of 3G networks in the United Kingdom. The auction process in 2000 raised the amazing sum of £22.47 billion.

Other countries followed the U.K. example. Germany raised even more money than the United Kingdom by auctioning bandwidth, but France and elsewhere raised considerably less due to bad timing and poorer processes. Some countries such as Sweden found the whole auction process too sordid and simply issued licenses at a fixed fee and a profit share. While it might be expected that the substantial cost of winning an auction in a territory such as the United Kingdom or Germany would affect pricing, there is little sign of this with the costs of mobile phone calls becoming increasingly affected by pan-European comparisons. While this process has been criticized in that it moved financial resources that might have been used to improve infrastructure into general government taxation revenue, it was used in later times for auctioning further bandwidths when they became available.

The evolution of the 3G protocol to include high-speed packet access (HPSA) in 2005 meant for the first-time mobile phones could achieve a respectable data download rate above 1 Mbps, which was approaching that achieved by many landlines at that time. This enabled the rapid expansion of mobile social media with the growth of Facebook and other apps we have learned to love.

While in the period from 2000 as 3G networks were being rolled out, there were competing regulatory standards in different territories, the market for handsets in particular had become so large that major suppliers such as Nokia and Samsung were prepared to invest in producing handsets to meet this demand.

3.2.4   Fourth Generation (4G): LTE, a Pure IP Packet-Based Solution

With 3G phones, digital packet switching of data had been used and, in later variants of the specifications, higher data rates were introduced. Packet switching mirrored the developments of fixed telephony where IP switching had become the dominant form of data transmission after 2000. Inevitably, it became clear that phone networks using an all-IP packet-based system for both speech and data would offer network operators and users advantages in both performance (data rates and capacity) and reduce the complexity (hence, cost) of networks. The 4G development, which started in 2005, was an inevitable consequence of the need for more bandwidth and adopted a pure IP packet approach akin to Wi-Fi.

Almost inevitably, two competing standards were conceived: WiMAX and Long-Term Evolution (LTE) with inevitable issues of noncompatibility. Eventually, the LTE path (which became synonymous but not completely aligned with 4G) became the route by which different communities were able to unite behind one international standard. The extensive work in standardization organizations plotted paths for long-term convergence of specifications and interoperability of phones in different areas. With it came substantially improved data rates that meant that applications such as mobile video transmission were practicable.

The convergence of specifications enabled by LTE made it easier for low-cost suppliers from Asia to enter the market for handsets and this, in turn, made it harder for established national players to survive as handset manufacturers [11].

3.2.5   Fifth Generation (5G): The Internet of Everything

5G, which is essentially an extension of LTE to support higher frequencies and higher speeds, with a somewhat updated internal architecture, is still being evolved and should probably be considered as a range of specifications covering a variety of transmission frequencies and data rates [12]. Although phone operators launched 5G services in 2019, they are only a small part of what will eventually embrace the Internet of Everything, leading to the prospect of just about everything in the world interconnecting (at least in theory). Major suppliers to this market, such as Qualcomm, clearly endorse this view of the future as witnessed in the report that they commissioned [13]. By 2020, the world began to wake up to the potential political, social, and economic implications of this development with the rather belated understanding that phone networks were a key strategic asset of society and as such who owns, controls, and uses them is of major importance, as highlighted in the first chapter of this book.

These five generations of specifications created a plethora of acronyms and an extremely complex evolution path whereby a mobile phone user could see ever-improving performance and roaming of their phones while remaining largely oblivious as to complexities of the developments that made this happen. Figure 3.6 is a diagram of the evolution path of these specifications.

While it displays a bewildering number of acronyms, the diagram does effectively show the key evolution paths involved. In particular, it illustrates how the GSM specification evolved through the LTE path to provide the basis for the current 5G specifications.

3.3 The International Rollout

Demonstrating the basic feasibility of cellular mobile telephony opened up an enormous market opportunity for the supply of equipment and services to support this new industry. Although existing PTTs and their supply base would appear to be the most likely players to benefit from this opportunity, in many cases newcomers to the industry were the main beneficiaries. While this was due to several issues relating to the political and commercial pressures in specific countries, one should also bear in mind the findings of Christensen [14] in his seminal work on the data disc storage industry that showed how incumbent market leaders can fail to benefit from innovation even if it affects their industry directly.

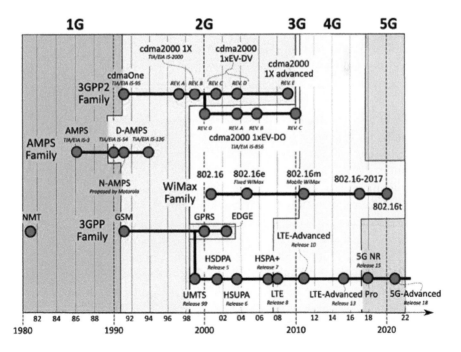

Figure 3.6 Evolution of mobile phone generations. (Source: Michael Bakni CC-SA 4.0.)

In two major markets, the United States and the United Kingdom, the point at which the opportunity began to become clear coincided with dramatic political changes that were to affect the telecoms industry.

3.3.1 The United States

As the birthplace of telephony, the United States could be expected to be the natural leader of mobile telephony. However, as the previous graph of the take-up of mobile telephony by country shows, the United States did not achieve a preeminent position in mobile telephony. In terms of take-up of telephony, it clearly lagged areas such as Scandinavia and some Far Eastern countries. It is also apparent that the U.S. supply industry made big gains only in some key areas, such as semiconductors and software. Other areas, such as handset provision and transmission equipment, were increasingly dominated by imports.

While there are several reasons behind this, one major factor was the breakup of the Bell telephone system during the 1980s. At a key period for rolling out cellular phone infrastructure, the United States was engaged in the fragmentation of the Bell monopoly in 1984 into "Baby Bells" (regional subsidiaries of Bell providing telephony services). The Bell system was a direct descendant of the network pioneered by Alexander Graham Bell after his invention of the phone in the 1880s. It had become one of the biggest companies in the world and was a fully integrated operation that not only provided telephone services but also largely made its own equipment and had enormous technical capabilities, witnessed by the invention of the transistor in Bell Labs in 1948 and much of the concept of cellular phones. However, its very size had led to criticism over its operation of what was a near-monopoly across the entire U.S. telephone market. The 1984 breakup was not a new idea; the sale of what became ITT and its international assets in 1925 was due to the same concerns.

Up to the breakup, the United States had largely led the field in developing the concept of cellular mobile telephony with the original research coming out of AT&T (e.g., Fluhr and Nussbaum) and early trials of cellular operation being run in the 1970s. However, at the very time that a massive rollout of mobile telephony infrastructure was appropriate, the Bell System was in the middle of being broken up and was thus not able to engage in a nationwide roll-out of a unified system nor would the FCC grant near-monopoly rights to Bell or any other national operator [15]. The vast U.S. market could be seen as several hundred cities with a surrounding area that was mainly rural. Acting on this notion, the FCC decided to fragment the installation of cellular communications and auction the provision of mobile services initially in 90 major metropolitan areas. Typically, two licenses would be granted, one to a local telephone company, often the local Baby Bell (once formed in 1984), and one to an existing radio common carrier (e.g., paging) operator. Not surprisingly, the process created an overwhelming number of applications. The main 30 metropolitan areas were chosen, but the balance of applications was chosen by lottery. This process was then extended to the issue of licenses for secondary areas. Inevitably, this approach led to many applications from start-up operations of varying capabilities. Once franchises were awarded, funding and operational issues led to many mergers. Initially, roaming was difficult due to the residual effects of the 1984 restructuring and the commercial policies of carriers [16]. These issues prevented mobile telephony being effectively promoted in the United States in the 1980s, so the country lost its technical and commercial lead, and, even by 2008, it was still behind many other developed countries in terms of mobile phone market penetration.

Other than the early 800-MHz analog cellular band, regulators did not specify which protocols could be used on which band; this was left to carriers to decide for themselves. Carriers were largely divided between the GSM camp and the CDMA camp through 3G, but all accepted LTE for 4G and gradually restructured their spectrum for it.

3.3.2   The United Kingdom

In the United Kingdom, the rollout of mobile telephony was also heavily affected by the political climate of the 1980s. However, in this case, it can be argued that the effect of this climate was beneficial at least from the viewpoint of the consumer. The privatization of BT in 1984 was one of the flagship policies of the Conservative government led by Margaret Thatcher and, during the 1980s, it was expected that telecoms would become a competitive market with at least two national suppliers but with the "temporary" oversight of a regulation authority, originally Oftel, which later became Ofcom.

When confronted by the mobile phone opportunity, although the initial market was overseen by several government departments, the overreaching government demand of creating competition affected all decision-making. The early decisions on mobile phones were taken at the same time as the key discussions relating to the privatization of BT in 1984 [17]. It licensed two networks at the end of 1982:

1. *Cellnet:* A joint venture between BT and Securicor (a U.K. security company).

2. *Vodafone:* A consortium led by a relatively small U.K. defense communications company called Racal. They were in partnership with a U.S. operation Millicom, which had early experience of U.S. cellular radio systems in North Carolina.

What was remarkable was how little discussion took place on this decision. While the choice of the two operators was stated to be the result of an open competition that was independently judged, the wording of the government announcement notes that Racal provided the clearest plan for a rollout of a network with a significant commitment to investment and coverage. Presumably, the Cellnet choice was to pacify BT as the incumbent telephony operator. However, the government did not want BT to have complete control over Cellnet, hence the unlikely pairing with Securicor (a provider of burglar alarm monitoring and other security services) as an equal partner in Cellnet. This uneven partnership survived until 1999 when BT bought out Securicor's then-minority interest in the business.

The most surprising omission was that of any existing U.K. telecoms supplier, particularly GEC, which had ample funds to support this development and considerable technical resources in radio communications. It can be assumed that at the time GEC and other players were absorbed in the issues relating to the creation of BT and the development of System X. Anecdotal evidence states that Arnold Weinstock, the powerful CEO of GEC, "did not like mobile phones." He much preferred markets where the initial investment came from government with GEC often achieving a near-monopoly supply situation (such as with many defense projects), as described in my interview with Peter Rowley (see Section 5.4).

The two chosen suppliers and the government collaborated to agree on a standard for the networks. In line with the then-Thatcherite mindset, a best-of-breed standard was adopted from the United States, adapted to the 900-MHz band to be used in the United Kingdom, and called AMPS. The government was in a hurry to get the networks running with the first calls to be made in 1985 only 3 years after choosing the operators. Because of this, the U.K. supply industry had little chance to develop products for this new market and most of the equipment purchased came from established international suppliers.

The roll out of the two networks went remarkably smoothly with the first mobile phone call attributed to Ernie Harrison, the chairman of Vodafone, on New Year's Eve 1984 [18]. Vodafone led the way in take-up of phones and was able to claim in August 1988 to have 211,500 subscribers, 55% of all U.K. subscribers [19]. The development of the network was largely implemented by licensing telecoms infrastructure capacity from BT together with importing specialist communications equipment from major world suppliers, notably Motorola and Ericsson (both of whom had built up their U.K. capability at this time).

By the end of the 1980s, the market was growing steadily mainly based on business users and the bandwidth available for mobile telephony was beginning to fill up. Meanwhile, the U.K. government had decided to increase competition by introducing new operators. Bids were invited in 1989 for what was initially described as personal communications networks (PCNs) with three licenses offered. While this was seen as an alternative to the GSM standard being delivered in mainland Europe, by the time the networks were put into operation, the standards had merged.

Because all the original 900-MHz band for mobile phones was taken up, a new band at 1,800 MHz was made available for the PCN operators. The United Kingdom successfully lobbied that this frequency be subsequentially incorporated within the GSM standard. While existing telecom suppliers, such as GEC (particularly its Marconi subsidiary) and Plessey, expressed an interest and indeed had been involved in the initial development of the GSM standard, they were shunned, and eventually new licenses were given to two new consortia. After many changes in consortia membership, these emerged as:

- Orange (a new operator backed by Hutchison, a Hong Kong-based conglomerate, and British Aerospace);
- One2One (backed by Cable & Wireless and US West).

It would seem that the main criteria for choosing the two additional network operators were their perceived commitment and resources to invest the £100 million seen as needed to roll out a new network across the United Kingdom. The four operators developed parallel networks with little use of common facilities other than using much of the BT landline network as a backbone to their systems.

The extra competition and capacity that this development produced propelled the United Kingdom to become (for a while) Europe's biggest market for mobile phones. While the United Kingdom's manufacturing strengths may be questionable, the strength of its retailing sector is very impressive. The mobile phone market produced a plethora of retailers offering a wide range of offerings linked back to the mobile networks. A notable example of this was the pay-as-you-go tariff that offered a very cheap handset and a tariff based on usage that was collected by guaranteed credit card or bank payment. The dramatic growth in the U.K. market can be seen in Table 3.2, which is extracted from the market leader, Vodafone's prospectus and annual accounts.

The new operators commenced operations in 1993 with the existing Vodafone and Cellnet networks continuing with their analog networks. However, by the mid-1990s, all four networks were digital and operating under the umbrella of the agreed GSM standard.

The U.K. government used this expansion to enhance its taxation revenues by auctioning bandwidth to existing and incoming service providers. This was facilitated by the 1998 Wireless and Telegraphy Act and a complex and keenly fought auction was arranged for what was to become five licensed

Table 3.2
U.K. Mobile Phone Uptake

	1988	1991	2000
Total U.K. Subscribers	383,000	1,267,000	27,500,000
U.K. Vodafone Users	211,000	710,000	8,800,000
Vodafone Market Share	55%	56%	32%

operators of 3G networks in the United Kingdom [16]. The auction process in 2000 raised the amazing sum of £22.47 billion, with the five "winners," each paying between £4 billion and £6 billion being:

1. *Vodafone-Aircall,* the market leader.
2. *BT Cellnet,* the number two.
3. *Orange,* at this time owned by Vodafone (who had acquired it as part of its acquisition of the German Mannesmann company). Vodafone were required to sell it off as part of the authorization of the Mannesmann deal and it was shortly to be acquired by France Telecom.
4. *One2One,* by then owned by Deutsche Telecoms.
5. *TIW,* a U.S.-Canadian company backed by Hutchison Wampoa of Hong Kong. This was a new entrant to the public networks although it owned the secondary U.K. network Dolphin, which was used by emergency services. This would later become a struggling fifth network "Three" owned by Hutchinson.

While doing wonders for U.K. government finances, the cash drain on the winners undoubtedly restricted their ability to roll out the new services they envisaged, and a string of mergers and sell-offs ensued. Except for Vodafone, all the other telephone networks went through changes of ownership, some several times. To summarize the sometimes bewildering changes:

- Cellnet became O2. This was owned by BT until after a strategic review (debt reduction program) it "demerged" O2 and this was acquired in 2005 by Telefonica, the main Spanish operator [20]. In 2020, Telefonica merged its U.K. operation with the dominant cable operator, Virgin Media.

- Orange had a successful early run with a strong sales and marketing campaign that made it market leader in the mid-1990s, but this did not prevent it from changing ownership and ending up belonging to France

Telecom as a consequence of the Vodafone acquisition of its then-owner, Mannesmann of Germany. Later, the operational savings needed to fund the government 3G auction led to its merger with T-Mobile of Germany.

- One2One was initially the flagship of Cable and Wireless (C&W), the U.K. government's preferred competitor to BT. However, when C&W lost heart in the uphill struggle to compete with the incumbent BT's landline operation, they eventually sold One2One to Deutsche Telecom, where it emerged as T-Mobile, which was later to take over Orange.

- Three was always seen as the poor relation with a weak network outside the main urban areas. However, Hutchinson stuck with the investment and established the network as a serious competitor as part of their international ambitions.

3.3.3 Other European Countries

The main markets of France and Germany developed in a more predictable way initially with the national PTT taking the lead and main ownership of the initial operator. In France, the initial network was provided by Matra, a state-controlled military and telecoms conglomerate. In Germany, the mobile network was a product of collaboration between the German PTT, Deutsche Bundespost, and Siemens, the large electrical conglomerate. However, soon some competition was considered necessary, and a second operator was introduced, in the case of France, Société Francaise du Radiotelephone (SFR), and, in the case of Germany, Mannesmann.

Similar paths were followed in Italy and Spain, but the Nordic companies achieved a more competitive environment.

3.3.4 Japan

Early innovation had been a feature of the Japanese market with early networks introduced by 1986. However, an excess of liberalization meant that several competing protocols were in use with all the resultant incompatibility. This era was also one where there was a major commercial dispute between the United States and Japan about U.S. access of technological products to the Japanese market. One facet of this was an epic battle led by Motorola of the United States to gain access to the rapidly growing cellular market. After facing a long rear-guard action, Motorola gained this full access in 1990, by which time much of its technological lead had been eroded [21]. This urge for the creation of their own standards lasted throughout the 1990s. Three consortia were licensed in the 1990s, led by NTT, Nissan, and Japan Telecom. This led to a rapid growth in demand. However, the evolution of the GSM standard meant possible handset manufacturers such as Sony and Panasonic struggled to compete on the world market as they had weaker access to the technology than Ericsson and Nokia.

Prior to the universal adoption of the internet as a standard for data communications, Japan was very innovative in creating early applications for mobile data. A proprietary system called i-mode achieved tens of millions of users prior to 2000 who were able to access commercial sites for information on products and services [22]. While this was perhaps the first major achievement of mass mobile data applications for mobile phones, its failure to find users outside Japan meant that it would eventually be overtaken by the World Wide Web.

3.4　Suppliers to the Market

This major new market, in turn, created large supply opportunities to equipment suppliers in three main component areas: handsets, phone masts, and connection infrastructure.

Handsets are the most obvious opportunity. Every user needs one and this provides a unique opportunity to access a billion-dollar-plus mass market. A handset must be easy to use and relatively cheap (when its functionality is considered). This is similar to consumer goods such as personal radios that Asian suppliers had come to dominate. An overview of the handset units shipped is shown below, listing the main manufacturers at the time. Table 3.3 shows the explosive growth in worldwide demand with sales increasing a hundredfold in

Table 3.3
Mobile Phone Suppliers

Million Handsets	Country	1992	2000	2008	2018
Apple	United States/China	—	—	12	219
Ericsson/Sony-Ericsson	Sweden/Japan	—	41	93	—
LG	Korea	—	—	102	40
Motorola	United States	4	60	58	—
NEC	Japan	2	—	—	—
Nokia	Finland	3	126	441	—
Samsung	Korea	—	20	235	295
Siemens	Germany	—	26	—	—
Others		3	139	299	842
Total		12	412	1,222	1,540

This is my personal estimate using data from a number of sources, including Gartner and Wikipedia reports.

16 years. However, it also shows how handset suppliers found it difficult to hold on to market share in the rapidly evolving world market.

Until well after 2000, Asian suppliers were not able to dominate this market. This is due to the complexity of design and the protocols used in the phones. In practice, manufacturers close to key markets in the United States and Europe dominated the industry. In particular, the success of GSM phones meant that Nokia became for a time the dominant manufacturer in the world despite (or because) it only had a small home market in Finland. Other major manufacturers were Motorola in the United States and Ericsson of Sweden who formed a joined venture in 2001 with the Japanese Sony Electronics Corporation to make phones under the Sony-Ericsson brand.

There was no shortage of innovative U.K. start-ups to address the growing cellular market. One example is Technophone. This was founded in 1984 and established a strong reputation for supplying small mobile handsets including what is claimed to be the first phone to fit in a pocket. Its private shareholders eventually sold the company to Nokia in 1991 who were able to use the technology to drive it to the position of the world's leading mobile phone supplier in the 2000s (taking this position from Motorola).

The surprising volatility of the market may be shown by looking at the worldwide sales up to 2018. As Table 3.3 shows, the rapidly expanding market represented a high-risk/high-reward marketplace for suppliers. The rapidly evolving technology and enormous pressure on price meant that no supplier seems able to maintain its position for long. Even Motorola and Nokia, the leading suppliers in 2000 are now effectively no longer major handset suppliers (their mobile phone businesses having been sold off). In 2020, smartphone manufacturers Apple and Samsung had a strong position but were being overtaken by Chinese suppliers (notably Huawei) that are in the "others" row in the table. Increasingly, the hardware platform is commoditized, and all the product differentiation is in the software, both the operating system (dominated by Google and Apple) and the social media applications.

Base stations (and associated radio equipment) are complex and used in much lower volumes than handsets. Therefore, it is not surprising that early suppliers such as Ericsson and Motorola could dominate this area. The speed of the take-up of mobile phones made later entry into the market very difficult. Thus, while the U.K. manufacturers had the inherent technical capability to develop this equipment, they had neither the time nor a protected local market in which to develop an offering.

An example of how the politically driven process of rolling out a mobile phone network worked can be seen in the United Kingdom. Once a network operator had been selected, it was under enormous pressure to install, commission, and start operating its network. No one would buy a mobile phone unless it offered good coverage throughout the territory. This, in turn, meant the mobile phone operator had to choose established and proven products for its networks. In Europe, this gave a large advantage to Scandinavian suppliers who had been actively engaged in rolling out their domestic network before other European countries seriously engaged. This can be seen in the Vodafone prospectus in 1988 when it was demerged from Racal. Between 1983 and 1988, Vodafone spent £62.5 million with Ericsson, representing over 40% of the entire fixed capital investment in the network [19]. However, Vodafone was certainly prepared to invest in the development of products that it could not source on the world market. An example of this is the software for planning cells in its expanding network, which it used to gain a significant advantage over Cellnet in the early rollout of its network [23].

The political bidding system adopted in many territories for the allocation of mobile network licenses greatly favored suppliers of existing products. As seen above, these were often suppliers that had already pioneered equipment in their domestic territories. Suppliers of similar communications products such as Motorola were also encouraged to expand their facilities in the United Kingdom and provide equipment to the growing cell phone networks. Thus, in 2000, Motorola announced that, together with Cisco, it had received the world's first order for supplying a GPRS cellular network from BT Cellnet.

In connection infrastructure, it is often overlooked that a large part of mobile communications takes place through conventional landlines that interlink base stations with the associated call management and switching functions. This gives indigenous landline operators a strong opportunity and in most major markets the phone operators have not tried to build their own network, preferring to concentrate their resources on phone masts and marketing. However, for existing landline suppliers to adapt a network for mobile communications requires several additional capabilities beyond the electronic exchanges described in the previous section (such as a registry of phones). Again, suppliers with a strong link with local phone operators could develop these capabilities for local use before offering them to the world market. In the United Kingdom's case GPT (the dominant U.K. supplier in the 1990s) was actively discouraged by its shareholder Siemens from making such developments. This was also confirmed in my interview with Peter Rowley (see Section 5.4).

Thus, for the United Kingdom and the United States, the development of mobile phones provided immense benefits for their society, but their telecoms supply industries failed to see commensurate long-term commercial benefits from this enormous market opportunity.

References

[1]   Carding, N., "Pandemic Spurs NHS to 'Urgently' Speed Up Elimination of Pagers," *Health Services Journal*, May 11, 2020.

[2] Tuttlebee, W., "Cordless Personal Communications," *IEEE Communications Magazine*, 1992, pp. 42–53.

[3] "Amos E. Joel Jr., Cellphone Pioneer, Dies at 90," *New York Times*, October 27, 2008.

[4] Valdar, A., *Understanding Telecommunications Networks*, London: IET, 2006.

[5] Fluhr, Z., and E. Nussbaum, "Switching Plan for a Cellular Mobile Telephone System," *IEEE Transactions on Communications*, Vol. 21, No. 11, 1973, p. 1281–1286.

[6] "Father of the Cellphone," *The Economist*, June 6, 2009, p. 29.

[7] Agar, J., *Constant Touch: A Brief History of the Mobile Phone*, Cambridge, U.K.: Icon Books, 2003, p. 94.

[8] Thatcher, M., *The Politics of Telecommunications: National Institutions, Convergences, and Change in Britain and France*, Table 42, Oxford, U.K.: Oxford University Press, 2000.

[9] Temple, S., *Inside the Mobile Revolution*, 2010.

[10] GSMA, "History of GSM," 2017, https://www.gsma.com/aboutus/history.

[11] "Your Phone on Steroids," *The Economist*, May 30, 2015.

[12] "The Big Mobile-Phone Reset," *The Economist*, September 7, 2013, p. 64.

[13] Qualcomm/Omdia, "The 5G Economy in a Post-COVID-19 Era," 2020.

[14] Christensen, C. M., "The Rigid Disk Drive Industry: A History of Commercial and Technological Turbulence," *The Harvard Business History Review*, 1993, pp. 531–588.

[15] Agar, J., *Constant Touch: A Brief History of the Mobile Phone*, Cambridge, U.K.: Icon Books, 2003, Ch. 3.

[16] Spufford, F., *Backroom Boys*, London: Faber & Faber, 2003.

[17] *Report of UK Parliamentary Proceedings*, London: Hansard, December 16, 1982.

[18] Wallop, H., "The Battle Behind Britain's First Mobile Phone Call," *Daily Telegraph*, December 27, 2014.

[19] "Vodafone Prospectus," *London Times*, October 13, 1988, pp. 33–36.

[20] "UK Mobile Phone Auction Nets Billions," *BBC News*, April 27, 2000.

[21] Tyson, L. D., *Who's Bashing Whom? Trade Conflict in High-Technology Industries*, Washington, D.C.: Institute for International Economics, 1993, pp. 66–71.

[22] Agar, J., *Constant Touch: A Brief History of the Mobile Phone*, Cambridge, U.K.: Icon Books, 2003, p. 96.

[23] Standage, T., *The Victorian Internet*, London: Weidenfeld and Nicolson, 1998.

4

Datacoms

Datacoms is not a new concept (I will use this shortened version of data communications extensively in this chapter). It is technically part of telecoms, but the word telecoms is often used just as a shorthand for the transmission of audio speech over a telephony network. Speech telecommunication required the development of electrical technology to develop whereas datacoms could be carried out at much earlier dates using optical transmission systems such as semaphore signaling (see Section 1.6). Figure 4.1 reminds us that datacoms go back a long way. In virtually all early datacoms systems, the data sent was in the form of text with individual letters coded in a format that could be easily generated and decoded, such as semaphore.

While optical semaphore systems and their like were admirably simple and could achieve impressive transmission speeds (when compared to a dispatch rider), they had their problems, being labor-intensive and prone to the influence of weather and geography. Thus, early experiments in electrical transmission at the beginning of the nineteenth century rapidly led to the development of electrical telegraphy.

The invention of the transistor not only enabled the rapid improvement of fixed telephony (Chapter 2) and the development of mobile telephony (Chapter 3), but it also drove key developments in datacoms that produced today's Information Age with the internet, social media, and the almost instant availability of information throughout the world.

Appendix D shows the key timeline events for datacoms, and these are reviewed in the following sections.

Figure 4.1 Smoke signals, an early datacoms technology. (Source: John Pritchett.)

4.1 Telegraphy

The development of telegraphy [1] followed the discovery of electricity with some experimental developments even predating the Victorian age. By digitizing individual letters, text messages could be sent over long distances. Early telegraphy was a manual system with an operator at each end of the transmission wire using simple codes to send a text message. Telegraphy had several inventors. In the United Kingdom, Cooke and Wheatstone, working in an uneasy partnership, developed and patented (in 1837) a rather complex system using five indicator needles driven by five pairs of wires. At the same time, Samuel Morse was developing a much simpler system in the United States. This used a single wire with information sent as a series of short and long pulses of electricity. Alfred Vail developed his own code with each letter having up to four short or long pulses. In a superb piece of pragmatic development with his partner, Vail counted the number of letters in a printer's setup and usually assigned the shortest codes to the most common letters. Thus, "e" was a single short pulse. Morse's development was demonstrated and patented by 1838 with the first Morse code machine operating in 1844. Morse is widely celebrated as the inventor of the telegraph, as shown in Figure 4.2.

In both the United Kingdom and the United States, it took a while for society to adopt the invention of the telegraph. Many observers simply believed they were seeing some sort of elaborate hoax, but, by 1852, there were substantial telegraph networks in many countries. The United Kingdom had 2,000 miles, but, unlike many countries, was slow to adopt Morse code keeping with the earlier developed needle telegraph of Cooke and Wheatstone. The telegraphy network was largely rolled out along the world's rapidly expanding railway network, which needed its own telegraphy network to control its operations.

In the United States, the telegraph network became dominated by Western Union, a private company that provided the first coast-to-coast connection in 1861. In the United Kingdom the early network was dominated by the Electric Telegraph Company, which had acquired Cooke's patents. This company and its competitors were acquired by the British Post Office in 1869. In most developed countries, separate telegraph networks were introduced for commercial and governmental requirements. Transmitting and receiving of telegraphs were done manually by elite operators working in large telegraph offices that could achieve a transmission rate of up to 30 words per minute. Interconnecting offices involved writing messages and then retransmitting them between offices. Demand for telegraphy grew dramatically from 1850 with inevitable delays at busy offices. In some countries, notably France, message transmission

Figure 4.2 Statue of William F. B. Morse, inventor of the telegraph, Central Park, New York, erected 1871. (Source: janoma.cl is CC BY-SA 2.0.)

between local city offices was supplemented by a system of pneumatic tubes that could take multiple paper messages in a container rather than the need to retransmit over the relatively short distances.

The possibility of using this technology with undersea cables was soon identified, with the first cable linking England and France operating in 1851. A transatlantic cable was laid in 1858, but failed one month after completion. The scene was now set for what has been referred to as the "Victorian World Wide Web" with cables stretching from the United Kingdom as far as Australia by 1872. The benefits of almost instantaneous communications across the British Empire led to rapid linking of many countries by cable [1]. Figure 4.3 shows the *Great Eastern*, the biggest ship of its time, when laying an early Atlantic cable in 1865.

The increasing demands for telegraph traffic encouraged the development of automation for both receiving and sending messages. The Wheatstone Automatic Transmitter (a machine to transmit the Morse code message from a simple typed tape) was introduced in 1858. Stock tickers were available in the 1860s (see Figure 4.4) giving a direct printed record of stock prices. Since the 1870s, automation reduced both the dependence on skilled operators at each end of the telegraph line and speeded transmission. Other developments in this era included the introduction of simple electrical repeater stations to allow transmission over longer distances along with basic duplex and multiplexing techniques to make more efficient use of the wires.

Surprisingly, most telegraph networks operated independently of the telephony network that began to be rolled out at the end of the nineteenth century. In most countries, the local post office organization assumed responsibility for telegraphy, which was an adjunct to the postal service while pioneering telephony services were left to private companies. The thriving telegraph industry

Figure 4.3   Great Eastern laying Atlantic cable. (Source: Morphart Creation, Shutterstock ID 306898910.)

Figure 4.4 Edison ticker tape machine. (Source: James Steidl, Shutterstock ID 59320414.)

of the 1870s served as a training ground for inventors Thomas Edison and Alexander Graham Bell, whose inventions were to power the development of the telephone and other electrical devices. Teletype machines were developed to link the transmission and reception of telegraphs to typewriter operations.

However, by the 1970s, the telegraph system was largely seen as an anachronism with its dedicated dispatch riders and use for high-importance (or high-status) messages such as births, marriages, and deaths [2]. The last telegram was sent in the United Kingdom on September 30, 1982, after 139 years of service. However, this development was an interesting precursor to the development of the later World Wide Web and was of great importance in the development of politics and trade between 1840 and the 1950s.

4.2 Telex

With the establishment of the telephony, it could be seen that it would be possible to automate the transmission and switching of telegraphs as in the telephone network. Thus, the telex network came into being. This used the same concepts as telegraphy but largely automated the transmission and receiving process often by using punched tape as a medium for the storage of messages. Each letter of the message would be coded as a vertical line of punched holes across the paper tape. Early work was carried out in Germany on their railway network between 1926 and 1933 and established a network operating at 50 bps. Similar developments were carried out in the United States with a 45-bps system developed by AT&T and introduced in 1931 as the TWX (Telewriter Writer Exchange Service). Unlike most other industrialized countries, this service used the voice telephony network using early modems with provisions to prevent users switching between the two modes. In some developing countries, short-wave radio links were established by its PTT before a full wired network could be established. An error-correcting protocol was developed to enable telex messages to be transmitted over radio (TOR), thus enabling telex messaging to remote areas.

The inherent nature of a telex message meant that it could be stored on paper tape and onward transmitted later if the line was busy or if the message needed to cross into a different network. Later developments involved the introduction of standard codes to allow interoperability of networks and multiplexing of signals. Electrical switching like that used in telephony began to be introduced from the 1930s. The United Kingdom was late in introducing automatic switching but introduced Strowger-based automatic exchanges from the 1950s. By the 1960s, there was a worldwide network of telex machines (printing and transmitting terminals), linking many organizations with a speed of 45 bps.

The telex was often a central part of an organization's communications network (see Figure 4.5). By the 1960s, the telex process was well established, with telexes being prepared using punched tape, each vertical line of the tape corresponding to one letter using the 5-level Baudot code. AT&T's TWX network adopted the 7-level American Standard Code for Information Interchange (ASCII). While providing what is seen now as a slow data transmission rate, the offline preparation of messages on punched tape enabled a high level of utilization of the network. Even today the system continues to be in use in specialized applications (such as maritime operations). While relatively slow, the telex system had several key advantages:

- Each transmission had a key identifier data tag that meant the recipient could be confident of the identity of the sender. Similarly, for the message to be sent, an answer-back from the target address (unlike the internet) was required.

- The system was simple and robust using well-established processes such as punched tape.

- Although individual countries or, in some cases, companies could have proprietary networks, it was feasible to create gateways for these networks to interoperate.

- The system maintained a reputation for commercial integrity that made banks and financial institutions long-term users until satisfactory alter-

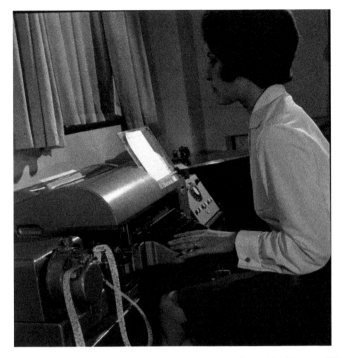

Figure 4.5 The telex, a vital part of most organizations up to the 1980s. (Source: BT Archives.)

natives could be established (such as facsimile and the Society for World-wide Interbank Financial Telecommunication (SWIFT) interbank network).

It was only with the introduction of the fax (facsimile technology) over the telephone network in the 1970s that this network became obsolete for many applications. However, the robustness and worldwide coverage of the telex system led to it lasting well into the twenty-first century. Some organizations even set up gateways to transfer messages between e-mails and telex. As worldwide coverage of e-mails was established, the separate telex systems were switched off. BT ceased telex operations in 2008. However, the short-wave TOR system was still used in maritime operations well into the 2020s.

4.3   Data over Voice Lines

With the development of computing from the 1950s, it did not take long for people to start to explore the possibility of sending computer data over the telephone network. Doing this required a modem (short for modulator/demodulator), which would turn data into audio signals. The voice phone network is engineered to transmit speech at frequencies between 300 Hz and 3,400 Hz. Thus, any data transmission must operate within these limits. A primitive form of modem (voice-coupler) predated the invention of the computer and was used in the telex network from the 1920s and as part of early systems to send images through phone lines for newspapers and other applications.

As the leader in the introduction of computers, the United States led the way in the transmission of data over phone lines. Data over voice lines began to be extensively explored by the U.S. military and others during World War II for sending encrypted voice messages and, by 1959, direct coupling of terminals to phone lines was possible in the United States, as Bell had released its Bell 101 modem initially designed for use in a military application. This modem ran at a pedestrian 110 bps, which was faster than the telex network. In line with its monopolistic situation, the U.S. Bell network prevented other suppliers offering products. These early modems continued to use a simple frequency shift keying (FSK) with acoustic coupling feeding into a standard Western Electric telephone (see Figure 4.6). Data speeds were progressively increased by the introduction of the Bell 201 modem (2,400-bps half-duplex) in 1962. However, an important (and overdue) decision by the FCC in 1968 (the Carterfone decision) allowed third-party suppliers to offer competing devices provided that they also used an acoustic coupler or a protective coupling device. Other suppliers, notably Vadic (part of the U.K. Racal group; see Section 5.4), developed modems operating at speeds up to 2,400 bps, but suppliers faced increasing difficulty in working with the Bell network. The FCC in the mid-1970s allowed direct connection of modems to the network, provided the modem met a stringent AT&T set of tests.

Figure 4.6   An early modem with acoustic telephone coupling. (Source: Doug McLean, Shutterstock ID 1109061452.)

With the progressive liberalization of supply for modems and the increasing use of microcomputers, there was considerable innovation in both the speed and functionality of modems. The Hayes standard was introduced by a supplier of that name, which provided for standardized functions such as a dialing; this enabled external modems to connect to microcomputers using a standard RS 232 port allowing the computer, rather than the user, to dial the telephone. Meanwhile, a series of standards were established to enable progressively faster data transmission, notably V21 (300 bps), V22 (1,200 bps), and V22bis (2,400 bps). A series of new modulation techniques enabled data rates well beyond those initially envisaged. This progression is set out in Table 4.1.

While the United States led the way in data communications, other countries were less hampered by regulation. In the United Kingdom, for instance, several start-up companies began to supply modems to meet the growing demand from businesses and individual consumers. An example is Pace Microtechnology PLC, which started life in 1982 and went on to supply a range of modems to users in the United Kingdom, United States, and elsewhere. As the market evolved, its product offering expanded to cover set-top boxes for satellite TV and built itself up into a significantly quoted U.K. company valued at over £1 billion before being acquired by Arris, a U.S. corporation [3]. What is notable is that, although the telecoms established suppliers in the United Kingdom and elsewhere were technically well equipped to address this new market opportunity, most of the business was taken by start-up companies that were much more capable at addressing the marketing issues that came with the new technology.

Table 4.1
Main Modem Standards

Connection Standard	Modulation Type	Bit Rate (kbps)	Year Released
Bell 101 modem	FSK	0.1	1958
Bell 103 or V.21	FSK	0.3	1962
Bell 202	FSK	1.2	1976
Bell 212A or V.22	QPSK	1.2	1980
V.22bis	QAM	2.4	1984
V.27ter	PSK	4.8	1984
V.32	QAM	9.6	1984
V.32bis	QAM+trellis	14.4	1991
V.32terbo	QAM+trellis	19.2	1993
V.34	QAM+trellis	28.8	1994
V.90	Digital+QAM	56.0/33.6	1998
V.92	Digital+QAM	56.0/48.0	2000

The telephone industry did not greatly adapt its offering to accommodate data transmission until the arrival of ISDN. Most major organizations did not want to face the issues of making connections over dial-up lines. Direct connection of lines was an option to an organization that was prepared to pay the considerable cost of line rental. This approach thus became widely adopted for data communications in critical applications such as telemetry for control and monitoring of widespread equipment (as in the utilities) and the interconnection of geographically dispersed organizations.

4.4   Facsimile/Fax

In most cases, datacoms was used exclusively for text transmission; however, the developing technology also enabled the concept of transmission of facsimile information over phone lines. This has a major effect on communications starting in the 1980s. However, the concept of facsimile transmission over phone lines was not new, a patent for a primitive version of a facsimile was filed in 1846 by a Scottish inventor Alexander Bain, who used a telegraph wire to link 2 pendulums. Subsequent developments enabled the transmission of pictures by telegraph of wanted criminals to be carried out in the early 1900s. These were only used on high-value applications, one being the transmission of photographs for the newspaper industry, which had been demonstrated by AT&T in 1924. However, transmission rates were slow and the equipment that was used in low quantities was prohibitively expensive for most commercial applications.

The dramatic growth in use of fax machines was largely a by-product of the development of document photocopying. Xerox which had key photocopying patents also patented what is regarded as the first modern fax machine in 1964. However, these machines were still bulky and expensive, using similar technology to photocopiers. Documents are scanned, which, in turn, generates a phone signal. The entry of Japanese competitors in the 1970s drove down the cost of such machines while greatly increasing their ease of use. The development of facsimile transmission was an attractive development for the Japanese market as their pictorial script has many more characters than western languages. The development of standard interfaces and protocols for fax machines also simplified the entry of third-party manufacturers into the market (Figure 4.7).

The Carterfone decision of 1968 in the United States enabled independent supply of fax machines and the CCITT (the international telephone standards organization) agreed to a series of standards that enabled interworking between different manufacturers' machines. These competitors were usually entering the market on the back of their sales of photocopiers, with which the fax machines shared many features (such as a large appetite for paper). Every office could have its own fax machine with the consequent decline in the use of

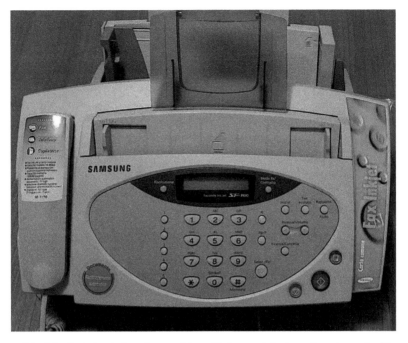

Figure 4.7 A typical stand-alone fax machine with its own telephone for voice calls. (Source: Pittigrilli, CC BY-SA 2.0.)

telex. The commercial machines were designed to fit into the existing telephone network and included dialing capabilities to use the voice network. Invariably, the fax machine had its own dedicated telephone line and most companies' letterheads included the fax number as a vital piece of communications information. The use of a dedicated fax line suited the telephone network operators well and at its peak fax traffic accounted for around 13% of companies' phone bills [4]. By the 1980s, the fax machine had become almost as established an item of office equipment as the telephone, and a series of standards set by the CCITT allowed for faster transmission with faxes using digital scanning and transmission in line with the then-current modem transmission standards.

Worldwide sales of fax machines peaked around 2000 with the sale of 15 million units and then their sales started to fall as usage declined after the millennium like that of the telex, although not as rapidly [4]. The cause of this decline was the wide acceptance of desktop computing and the use of the internet to facilitate easy data transmission across the world. The international transmission standards on the internet made the stand-alone fax machine unnecessary with a computer able to fulfill the same function if it is connected to the internet. Images of documents could be easily sent as e-mail attachments, removing the need for the intermediate stage of printing required by faxes.

However, the very properties of the fax that weakened its market attraction (when compared with e-mail) also provide reasons for some organizations to continue its use. In regulatory or strongly protected activities (like banking and health care), its integrity helped to prevent computer fraud, allowing its usage to continue. For subscribers without a computer or printer, a simple fax machine continues to offer an easy way to transmit documents. In some situations, a plethora of incompatible computer systems can make the standardization on fax transmission the only sure way of transmitting images; thus, in 2017, the U.K. National Health Service (NHS) was reported to be the world's biggest customer for fax machines [5]. Despite a directive from a U.K. government minister to stop using this "archaic" form of communication, some usage continued beyond 2020.

4.5   Growth of Early Data Services

With the increasing speed of data communications came the early growth of data services. While it would take the development of the internet and the World Wide Web to provide the widespread services that we now see, early pioneers responded to the opportunity of using data over phone lines to provide basic services. The main barrier to data services being adopted by consumers was the need for each household to have a data terminal and a modem to access the service over their phone line: their adoption. This was a fundamental issue in the 1980s when home computing was largely a preserve of a small band of enthusiasts. However, in several countries, what were often referred to as videotex services were developed. In these, the existing phone network provided a gateway whereby basic online services could be accessed.

Perhaps the most remarkable example of an early service offering was provided by the French Minitel service (actually, it is correctly termed Teletel, but it widely known under the name of its Minitel terminal). This aimed to overcome the "chicken-and-egg" problems described above by providing households in France with a dedicated terminal (see Figure 4.8), a small black-and-white monitor, as a simple free loan (with a charge for phone time). The service was initiated in 1982, and its first offering was the provision of an online national telephone directory. Other services were rolled out that would later become familiar to internet users including mail order sales, online news, ticketing for trains and planes, gaming, and messaging. The service was widely adopted with over the following decade, with 1.6 million terminals in use in France by 1986 [6] and over 5 million terminals in 1989 [7].

A slightly different approach was adopted in the United Kingdom and some other countries where in 1974 TV receivers became available using the Ceefax system. In this system, information was broadcast within the television

Figure 4.8 Minitel black-and-white terminal. (Source: Tiem, CC BY-SA 3.0.)

transmission. The television thus acted as a passive one-directional information source offering several hundred pages of information. The system became widely used because it was available at little additional outlay and was continued until 2012, by which time it could be replaced by smart TV services. In some countries such as the Netherlands, a hybrid teletext system was evolved using an adapted TV receiver and a modem to access a Minitel-like service. In the United Kingdom, Prestel was launched in 1979 and required a computer to host the two-way data communications and consequently only achieve low usage before it was terminated by BT in 1996.

Other countries adopted Minitel-based services, including many European nations plus early trials in Canada and the United States. In most other countries, these services did not achieve wide usage mainly due to the reluctance of the telephone operator to subsidize the purchase or rental of the terminal. One exception was the Netherlands where the service achieved a reasonable level of usage. Despite the trials in North America, the Minitel service never achieved high adoption because of the reluctance of users to buy a special terminal. The growing user base of early proto-ISPs such a CompuServe and AOL, which used computer-based systems, proved to be a longer-lasting solution to the North American market need.

The French Minitel system continued with millions of users into the twenty-first century but the power of the internet gradually eroded its appeal, and it was eventually retired in 2012. Despite the large amount of funding and effort applied by France to the Minitel project, it is hard to discern any obvious benefit derived by the French online marketplace compared with other European countries such as Germany and Italy.

4.6   High-Volume Data Transmission

While impressive increases in data transmission rates had been achieved by modems transmitting data down a switched telephone line, it became clear that higher data transfer rates could be achieved by integrating the functions of voice switching and data transmission. Several alternative approaches were developed as the demand for data transmission increased.

The development of semiconductors led to a rapid growth in the computer industry from the 1960s. Whereas initially computers were regarded as large, self-contained items operating within a single organization, the benefits of interconnection soon became clear. This trend increased in the 1970s as the minicomputer was introduced by manufacturers such as DEC alongside the large mainframe computers from suppliers such as IBM. The ways of interconnecting several computers for the purposes of file-sharing or messaging began to be explored. This is usually described as networking. Military applications sought to use the telephone system using early modems as described earlier. However, many of the early developments of networking took place outside the realms of the telecoms industry on behalf of the military under the guise of local area networking (LAN) or wide area networking (WAN), with early products being developed by the computer industry or increasingly by new start-up companies.

Thus, during the 1970s and 1980s, increasingly sophisticated methods of computer networking were being offered using a variety of newly developed protocols and equipment. Examples of this are the IBM token ring network (first introduced in 1984) and the Ethernet, which was initially developed by Xerox and became standardized as IEEE802.3 in 1983. To this day, Ethernet remains the basis of both home and data-center computer networking.

While the detailed nature of computer networking is outside the scope of this book, it created a whole new industry with suppliers such as 3-Com and Cisco as suppliers to this market. They were able to move to challenge the traditional telecoms industry suppliers as the demand for datacoms on the telecom networks grew. In turn, as the volume of datacoms increased, organizations increasingly began to consider voice communication as a smaller part of the communication needs, which could be digitized and carried on their data network.

4.6.1   Dedicated Lines

The option of having dedicated phone lines has always been available to organizations. This would typically be used by large multisite organizations that were prepared to pay a premium for having a continual line between their locations. With the introduction of computers and the resulting data traffic, these dedicated lines could offer high data rates (up to several megabits per second), albeit at a considerable monthly rental cost. This capability enabled increasingly sophisticated interconnections between computers with the development of packet switching as a precursor to the establishment of the internet [8].

4.6.2 X25: Virtual Packet Switching

The evolution of packet switching is discussed in a later section. An early precursor of this major development in data transmission was the introduction of the X25 standard by CCITT in 1976. A simple schematic view is shown in Figure 4.9. This gave a way whereby data could be assembled into packets and routed efficiently through a telecoms network. This provided a more cost-effective way to provide data communications while fitting within the overall architecture of the existing telecoms networks. Throughout the 1980s, the increased demand for international data transmission led to the increasing development of the X25 standard as a way of providing data transmission though multiple networks in areas such as Europe [9].

The availability of virtual packet switching came along as many phone networks began to dramatically increase their transmission capacity between exchanges. In the United Kingdom, the Kilostream (64 Kbps) and the Megastream (2, 4, 34 Mbps) leased-line products offered higher data rates to address the growing market for datacoms. Initially, copper coax cables were used, but later fiber-optic connections were rolled out.

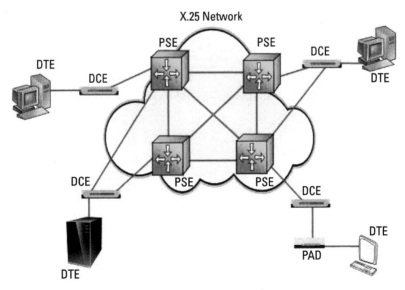

Figure 4.9   X25 schematic diagram. (Source: Adamantios.)

4.6.3   ISDN

The introduction of the ISDN heralded the beginning of the introduction of a network architecture more suited to data transmission with the normal telephone connection in the United Kingdom suddenly being able to support two 64-kbps voice or data channels for domestic users. The transmission was based on a series of protocols introduced by the CCITT in 1988 that allowed for an integrated approach to the transmission of voice, data, and video. Instead of data having to be converted to audio signals, audio telephone calls were digitized at source into a format compatible with onward transmission across the existing switched network, which had already digitized the transmission of voice between many switching systems.

There were two basic formats applied for ISDN. The basic rate of 2 times 64-kbps bearer (B) channels, plus a 16-kbps channel called the D channel for call control signaling and some low-speed data applications, over existing copper lines, targeted at domestic and small business users.

The primary rate made up of 30 bearer channels plus a 64-kbps D channel, providing a total transmission rate of up to 2 Mbps using copper twisted-wire pairs that were aimed at business users in conjunction (in the United Kingdom) with the Megastream connectivity.

ISDN gained wide acceptance in Europe, particularly for business users, but was only ever a niche product in North America. For the first time, data could be sent over the phone network without using a modem and the high quality of speech transfer made the network popular with broadcast operators. However, the relatively small gains in data rates offered with the basic rate transmission compared with the higher line charges by telephone operators in countries such as the United Kingdom did not make the technology overwhelmingly popular to domestic users in the United Kingdom and elsewhere.

4.6.4   Asymmetric Digital Subscriber Line

While ISDN was expected to be a long-term solution to the need for ever-increasing data rates to a telephone network's domestic and small business customers, it was rapidly overhauled by the asymmetric digital subscriber line (ADSL). BT has announced that it will not support ISDN beyond 2025. The technique of digital subscriber lines (DSLs) had been developed by Joseph Lechleider at Bellcore in the United States by 1988. He started by showing that it was possible to inject a higher-frequency carrier signal onto a standard copper telephone cable. This carrier could, in turn, carry data at a much higher rate in the downstream direction than achieved by the ISDN basic rate (over 1 Mbps, dependent on distance, compared with 128 Kbps with ISDN), while still supporting an analog telephone line on the same copper pair. The digital signal could then be separated from the audio at the exchange end of the consumer line with the two streams going on different paths through the network. While the issues of exchange-end termination of the ADSL signal required some sorting with the need to inject the high-frequency carrier signal into the consumer's transmission path after the exchange, the data stream could now be extracted as a standard process and passed on to independent internet service providers if required. In order to optimize performance for the mass-market application, such as video delivery or internet service, it was best that the data rate in one direction was lower than in the other, hence the A in ADSL.

However, the real driver for ADSL was that it addressed one of the key issues in telecoms: the last mile. All major established telephone networks have millions of users connected by separate wire links to each house. The telephone operators feared that, with the increasing bandwidth requirement of video-on-demand, cable TV operators offering (optical) fiber to the home could undermine their monopoly of these last-mile connections and thus attack their core business. ADSL provided a way of countering this threat and largely seeing off these potential competitors. While ADSL is not as fast as optical fiber, a later form, ADSL2 offering data rates up to 24 Mbps is enough for ultrahigh-definition video. BT introduced ADSL in 1998 and it became the main route by which consumers accessed the internet in the United Kingdom. The consumer interface usually included an ADSL gateway (sometimes still called a modem), together with a router to provide wired Ethernet and Wi-Fi transmission to devices within the premises. While fiber to the home remains a stated aspiration of most telephone networks, they rolled this out at a relatively leisurely rate, having seen off most of the competition from cable companies.

4.7 Packet Switching and the Internet

Traditional telephony systems as described in this chapter are based on the simple principle that a transmission path is first set up between the sender and the receiver. This path is then used for the duration of the message. Even with the introduction of digital transmissions across the telephone network, such as those in the previously described digital protocols, this principle (often described as connection-orientated transmission) still applied to data transmission. While the obvious way to communicate (first establish a path and then send the information), it suffers from two key issues:

1. The channel allocated to the message remains open throughout transmission and thus this method can be relatively inefficient if the information is being sent at an irregular rate and the transmission channel is held for exclusive use during the connection. Even on a simple voice

telephone call much of the time is spent in silence. However, the network invariably charges for the time allocated to this channel.

2. The communication is not robust in that, if the transmission path is disrupted, the channel must be reestablished. This is not usually a big issue with short voice calls, but becomes much more significant in data transmission.

In the 1960s, it had become apparent to military planners, particularly in the United States, that there was a need for robust communications to share data. Early work was commissioned by the U.S. Air Force and was carried out by Paul Baran at the RAND Corporation. They developed a concept that they described as distributed adaptive message block switching. Independently, the U.K. National Physical Laboratory (NPL) had been researching similar issues and, in 1966, Don Davis presented an article describing the similar but much more much more catchily named packet switching. Inevitably, the two parties were introduced to each other, and the resulting development produced Advanced Research Project Agency Network (ARPANET), which went live in 1969 and is generally accepted as the first major implementation of a packet switching wide area network.

The concept of packet switching is elegantly simple. Information is split into data packets and each packet includes information needed to reach its destination and be reassembled into the full information. Once the transmission protocol is established on a network these packets can then make their way from the sender to the receiver using whatever route is most suitable. Thus, data transmission became resilient to breaks in the network. Figure 4.10 is an illustration of the original packet switching network developed by NPL; using a single Honeywell computer, it allowed data packets from several users to be sent to different users. From this simple beginning, the concepts could be used to build a network covering an entire country.

The development of ARPANET showed the potential power of a packet switching network, but it was envisaged as a network with limited access for military and associated applications. However, in the 1970s, developers could see the potential for an international network that was not devoted to one organization's need. The French Cyclades network developed by Louis Pouzin showed the way by which a network could be established without a centralized control. Cyclades introduced connectionless packet switching, in which every packet contained the full destination address, and thus did not require a connection to be established ahead of time. It also introduced the concept now known as internetworking, in which a packet-switched network operates atop transmission facilities, which could be other providers' packet-switched networks, independent of the underlying networks. This, in turn, led to the work of Vint Cerf and Bob Kahn on transmission protocols [10], which led

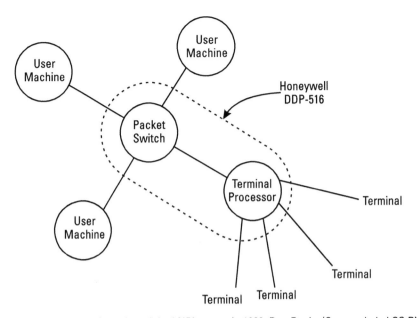

Figure 4.10   Diagram from the original NPL report in 1966, Don Davis. (Source: bakni CC BY-SA 4.0.)

to the development of the composite protocol Transmission Communication Protocol/Internet Protocol (TCP/IP), which is used as a basis for transmission on what is now known as the internet. This proved to be an extremely resilient piece of technology, which could be overlaid and integrated atop existing telecoms networks. A remarkable feature of this development has been the behind-the-scenes cooperation of many individuals and organizations to develop, adapt, and manage these protocols to the extent that worldwide internet communications is taken as a given feature of modern living.

Another important step in the development of public internet services was the establishment of the World Wide Web, arising from the pioneering work of Tim Berners-Lee at CERN in Switzerland published in 1989 [11]. This enabled information to be published using a standardized format of Handheld Device Markup Language (HDML) on a local computer and accessed throughout the world using newly conceived web browser tools [12]. This made the internet more accessible to the public at large, rather than being a tool best used by those more skilled in the use of computers.

Thus, by the 1990s, all the ingredients were in place for the development of a worldwide internet on which the World Wide Web alongside other internet services such as email could be accessed. The widespread deployment of Internet Protocol (IP) networks potentially made the classic switched telephone exchange obsolescent as its function could be replaced by digitizing speech and using routers to achieve all the necessary switching functions of the telephone network. Key innovations covered in Chapter 1, namely the development of semiconductor processors and the invention and subsequent deployment of fiber-optic communications, meant that the cost of transferring data was reduced by several orders of magnitude. The low cost of transferring information across a public internet that scaled up for high-volume data applications made it possible to send telephone calls at a lower cost than what legacy telephone carriers traditionally charged, especially for trunk and international calls. This substantially disrupted the business model of the telephone carriers and led to a migration towards voice over IP services. However, it is sobering to note that, 30 years later, most speech and much of the data are still handled by these obsolescent exchanges (BT plans to switch their System X exchanges off in 2025).

By 2000, the creation of the internet and the World Wide Web had begun to generate many business opportunities for products and services that are now considered fundamental. However, the development of social media was all but unimaginable and will not be considered in this book. The implications of the development of the internet and all that it brings is a major challenge for individuals, organizations, and society as a whole. The ways in which people access, process, and exchange data are changing rapidly, and the effects of this will require much research outside the scope of this book.

Enabling these changes are portions of the internet business ecosystem, which include:

- *Routers:* These are electronic devices that take data packets and transmit them onwards according to the IP. As the market developed, these products evolved to broadly two types: local and network routers. Local routers are used in most domestic and small business premises and primarily deal with local routing of data from ISP networks to individual users in the premises. Network routers manage the traffic flow within a communications network and between networks, often on a massive scale. Rather like the situation described with modems above, a whole new group of companies grew up to supply these items, notably Cisco in the United States, which was founded in 1984. Few of the established telecoms manufacturers pioneered the supply of routers, but they have increasingly found that their offerings have to include routing technology and companies founded on routing such as Cisco and Huawei of China have taken a large share of the market.

- *ISPs:* These provide access to the internet for their customers, and many provide the internet's backbone routing function. Some also offer hosting services for websites and e-mail accounts. Again, many of these organizations were start-ups as the market developed, but, in many territories, the local telephone network operator (such as BT in the United

Kingdom) offered ISP services and has remained a major supplier of these services. To provide internet services, an independent ISP must have access to the facilities of the national telecom service provider. In many territories, this relationship is policed by the regulatory body (Ofcom in the United Kingdom) to try to prevent the telecom supplier exploiting its monopoly position in charges for connectivity to end users.

Several ISPs entered the market having been active in providing consumer data communications prior to the evolution of the internet. A major example of this is AOL (originally called America Online), which started offering dial-up online services to computer gamers before evolving to offer broader online services. At its peak, AOL had 22 million subscribers in the 1990s. It made a spectacular (and expensive) merger with publisher Time Warner in 2001 before dramatically shrinking as demand changed, and it was eventually acquired by Verizon (a major U.S. wireline telephone and cellular operator) in 2005 [13].

The vertical integration and consolidation of the communications services sector after 2000 have led to having a smaller number of independent ISPs and, instead, a set of large companies that provide fixed and mobile telephony, internet access, and a range of associated data services.

• *Browser and web search engine software:* The power to view the content on the World Wide Web has always been crucial. Once the operation of a web browser had been demonstrated by Tim Berners-Lee in 1991, several generations of product were produced that had superior search capabilities. This was in response to a spectacular growth in the amount of information available on the World Wide Web. It is estimated that, by the early 2000s, there were over 1 billion separate pages on the web. This growth has continued driving the need to search out this information and efficiently display it. The actions of a search engine (which locates information against a search term) and a browser (which displays this information once it is located) are separate, but their marketing is often closely linked. In particular, the development of a pay-by-click business model for websites whereby advertisers would pay a small sum each time an advert is viewed by a user enabled these products to be developed and marketed at considerable cost while appearing to be free to the end user.

Search engines started mainly as a by-product of academic research. A plethora of start-up companies competed for users. After the 2000 dotcom crash, a major consolidation took place where many of these early search engines were subsumed into major brands such as Yahoo, Microsoft (its Live Search product has now become Bing), and Google.

With browsers, Netscape Navigator was launched in 1994 and gained market leadership. However, this position was fiercely attacked by Microsoft, which could bundle its Internet Explorer product within its Windows software suite (until challenged by regulatory authorities). Netscape was purchased by AOL in 1999; a derivative of its product, Firefox, has continued to be updated by the Mozilla Foundation. For a while, Microsoft then dominated the browser market until Google launched its Chrome browser in 2008 on the back of its strong position in search engines and became the dominant browser to the time of this writing.

4.8   Effect on the Telecoms Supply Industry

The above items are a small summary of a whole new industry that appeared from 1990 onwards and was to dramatically boost the demand for data communications. However, its dramatically different approach to revenue models largely bypassed the established telecoms industry.

While the telecoms supply industry had, in most cases, the technical capability to supply the equipment arising from the growth in datacoms, their limited experience of marketing products to end users made it difficult for them to quickly address the opportunities that arose. They also had a hard time pivoting to products that were based upon complex software that needed to cope with a rapidly changing internet environment. They were typically used to supplying to a few monopoly telecoms network operators, whereas many of the business opportunities were for products that needed to be sold directly to end users, both domestic consumers and commercial organizations. Historically, many telecoms network operators tried to maintain a monopoly of supplying end-user equipment. However, the U.S. Carterfone ruling of 1968 and the privatization of BT in the United Kingdom in 1984 made it difficult for the network operators to maintain a monopoly on end-user equipment supply that became increasingly available from different manufacturers through separate channels. This initially involved telephone handsets, but, as technology developed, this expanded to PBXs and into datacoms equipment including modems, routers, and fax machines.

Huge companies (such as Cisco in the United States) were spawned because of the boom in digital communications described here. Many were venture capital-funded as a major part of the California-centered venture capital industry that had already been established, funding businesses in semiconductors and life sciences. What most of the new businesses had in common was that they did not target the telecoms network operators as their primary source of revenue. They aimed either to make low-price, high-volume sales to end users or, even more radically, to offer a free-to-end-user product financed from advertising revenues.

The increasing globalization of the supply industry also allowed newly created Asian datacoms suppliers (notably Huawei) to enter the market for telecoms equipment in competition with established suppliers [14].

It is not surprising that the established suppliers to the telecoms industry largely missed these business opportunities. There were examples of companies offering products to compete in these markets. For instance, Japanese telecoms suppliers (such as Fuji) were already active in the supply of office equipment and were able to progress into fax machines via photocopiers.

The rapidly developing demand for digital communications did create demand from the telecom networks for new products to accommodate the vast growth in data transmissions, but the basic network structure proved remarkably robust in terms of its ability to handle an enormous growth in data traffic between 1950 and 2000. However, the new technologies introduced (particularly the adoption the TCP/IP packet switching) left the whole supply industry vulnerable to the commoditization of most of its hardware offerings and the resultant switch into becoming software system suppliers.

References

[1]   Standage, T., *The Victorian Internet*, London: Weidenfeld and Nicolson, 1998.

[2]   Carter Committee, *Report of the Post Office Review Committee, Cmnd. 6850*, London: HMSO, 1977.

[3]   Arris Corporation, "Arris to Acquire Pace plc for $2.1B in Stock and Cash," Suwanee, GA: Arris Corp Press Release, April 22, 2015.

[4]   "Last Gasp of the Fax Machine," *The Economist*, September 18, 2004.

[5]   "NHS Told to Ditch 'Absurd' Fax Machines," *BBC*, December 9, 2018.

[6]   Epstein, N., "Et voila! Le Minitel," *NY Times*, March 1996.

[7]   Thatcher, M., *The Politics of Telecommunications: National Institutions, Convergences, and Change in Britain and France, Table 38*, Oxford, U.K.: Oxford University Press, 2000.

[8]   Roberts, L. G., "The Evolution of Packet Switching," *Proc. of IEEE*, Vol. 66, No. 11, November 1978.

[9]   Steinmueller, D. A., "Economics of Compatibility Standards and Competition in Telecommunication Networks," *Information Economics and Policy*, 1994, pp. 217–241.

[10]   Kahn, C., "A Protocol for Packet Network Intercommunication," *IEEE Transactions on Communications*, Vol. 22, No. 5, 1974, pp. 637–648.

[11]   Berners-Lee, T., and M. Fischetti, *Weaving the Web: The Original Design and Ultimate Destiny of the World Wide Web by Its Inventor*, San Francisco, CA: Harper, 1999, p. 46.

[12]   Wu, T., *The Master Switch: The Rise and Fall of Information Empires,* New York: Random House, 2010, p. 282.

[13]   Hazlett, T. W., "A Lesson for Today's Tech Trustbusters," *Wall Street Journal,* January 10, 2020.

[14]   "The Company That Spooked the World," *The Economist,* August 4, 2012, p. 19.

[15]   Valdar, A., *Understanding Telecommunications Networks,* London: IET, 2006.

5

World Telecoms Supply

How does the U.K. industry compare with other countries? This book explores how the development of telecommunications in the United Kingdom interacted with suppliers in terms of creating wealth and employment. In Chapter 1, we saw that technology change is a major driver in the economies of developed countries and it is therefore important how the dramatic development of telecommunications in the period of interest affected the United Kingdom compared to other countries. Throughout the last 50 years of the twentieth century, all major developed countries faced the same challenges:

- Increasing wealth of the population and a growing internationalism, leading to a dramatic increase in demand for telecoms;

- The impact of major technological change creating new products and dramatically reducing the costs of telecoms;

- A concern to avoid monopolistic actions in the supply of telecoms services, leading to widespread privatization of the networks, attempts to inject competition into the industry, and the creation of independent regulation.

In this chapter, I will explore how the main developing countries responded to these challenges and how their telecoms industries fared as a result. This then gives a context in which we can review how the U.K. telecoms supply industry performed, which is the last section of this chapter. The trade performance of the countries is shown in Appendix A.

As the country that invented the telephone and often led the adoption of telecoms services, it is only right that we should start our review with the

United States. Here, from early in the twentieth century, a private company (AT&T) was a monopoly supplier of telephone services and the usage of telephones per head in the United States greatly exceeded those in other developed countries until the 1970s.

The review will then move on to major European countries where, at the start of the period, national PTTs controlled a monopoly on telephone service in each country. Moving to Japan, we will see that, after World War II (which led to a collapse of Japanese infrastructure), a state PTT resumed control in 1952. Finally, we will come to the United Kingdom, which started the 1950s with a European-style state monopoly PTT, namely the GPO.

5.1   North America

If one measures performance by the operating cost and efficiency of the telecoms system, the most effective approach (up to the 1970s) was that used in the United States where AT&T (often referred to as Ma Bell) not only was a monopoly operator of most of the telecoms network but was also the in-house supplier of most of the equipment (through its subsidiary Western Electric). It is difficult to compare the cost to the user of the U.S. system with those of other major economies, but it is widely believed that the AT&T monopoly was a very efficient supplier of telecoms services. Historically, domestic consumers had seen free local calls with higher charges for long-distance and international calls. This became a cultural fixture in the U.S. telecoms scheme (see Figure 5.1). The rapid growth of telephone usage in the United States was a curious hybrid between local innovation and national monopoly power [1]. The invention of the telephone by Alexander Graham Bell in 1876 (as described in Section 1.6) and the subsequent patenting took around 10 years to have a major impact across the nation. Bell had to win a struggle with the incumbent telegraph monopoly, Western Union, which came to see the telephone as an unwanted challenge to their business [2]. After the expiry of Bell's original patent in 1894, numerous innovative small organizations established telephone networks responding to local needs for communications in the period up to 1910. By then, the original Bell Telephone company had evolved from the initial local business into AT&T, which focused on providing long-distance connectivity. Locally based telephone companies were then absorbed by a mixture of coercion and incentivization into AT&T, led by its charismatic leader Theodore Vail with the backing of the powerful banker J. P. Morgan [3]. The monopolistic organization that they created even managed to acquire for a time Western Union (its telegraph competitor). In 1913, AT&T entered into the Kingsbury Commitment with the federal government whereby it agreed to divest Western Union and provide trunk call connectivity to independent local carriers. However, this

Figure 5.1 Ma Bell, part of the American dream (1957 advertisement). (Source: AT&T Archives and History Center.)

did not stop AT&T's monopolistic activities and it continued to be the subject of numerous disputes with the federal government over concerns such as pricing to users and the stifling of innovation. The U.S. government was able to dismantle monopolies that were against the public interest using antitrust legislation under the Sherman Act. However, AT&T was able to counter these concerns by pointing out that they operated what was widely regarded as the world's most successful telephone network. It was also a significant contributor to national defense with both research and infrastructure. This was backed up by an extremely close relationship with the federal government supported by an extensive lobbying network.

The concerns about how AT&T operated were well founded. Bell Labs was probably the best corporate research institute in the world in the period from 1930 to 1960 winning 7 Nobel Prizes and inventing the transistor, possibly the most significant invention of the twentieth century. However, inventions that did not match AT&T's vision for the development of the telephone network could be ruthlessly suppressed. An example of this is the development of magnetic recording (now used for tape and disc recording). This was pioneered at Bell Labs in the 1930s, but kept from publication because AT&T believed the ability to record speech would reduce the demand for telephony [4].

Similar arrogance was shown in the aggressive way AT&T fought to stop third parties from attaching anything to their network. This started with the rather ludicrous "Hush-A-Phone" case that took 8 years to resolve. Here Bell tried to stop a supplier offering a simple handset cover to keep telephone calls private. The 1956 judgment against Bell opened the door for the subsequent attachment of much more significant peripheral devices (although again strongly resisted by Bell). In 1968, the Carterfone Decision by the FCC opened the way for more third-party peripherals to be attached to the Bell network and the subsequent innovations seen strengthened the belief of many observers that the Bell monopoly was no longer in the national interest [5]. Thus, by the 1970s, there was a growing concern that the network was becoming stagnant and incapable of supporting innovation. There were several agreements made to accommodate the increasing use of computer connectivity and the provision of value-added services but these were unlikely to resolve the basic issue of AT&T's broad-reaching monopoly powers.

The breakup of the Bell monopoly in 1984 had a very significant effect on the United States' ability or willingness to compete in the increasingly global market for telecoms equipment. As described later in this chapter, the Bell system was not very involved in exporting outside North America. Its international arm had been spun off as the International Telephone & Telegraph (ITT) in 1925 so the Bell monopoly was only concerned with the issues of supplying and servicing the rapidly growing U.S. market for telecoms until 1984. The Bell system was broken up by AT&T divesting itself of its operating companies into 7 regional companies (usually called the Baby Bells) while it retained Western Electric, its development and manufacturing operation, and was allowed to sell into unregulated markets. AT&T had hoped to become a major player in the computer industry, but misunderstood that market and failed entirely to make a dent into it.

Chapter 3 showed how the United States failed to establish a technological lead in mobile telephony, although some suppliers (notably Motorola) actively used their strong technical base to develop export business for this new market opportunity. After many changes, many of the assets of Western Electric (the Bell manufacturing arm) and Bell Labs were assembled into a new company called Lucent in 1996. In the late 1990s, Lucent was the biggest telecoms manufacturer in the world and began serious efforts to market its products outside the United States. This included a large joint venture with the Dutch Philips group, which aimed to gain a significant share of the mobile phone handset market, but this was abandoned at great cost after only a year. Despite the increasing penetration of non-U.S. companies (such as Nortel and Ericsson) into the U.S. market, the Lucent company was recording rapid sales growth thanks to increasing demand from independent companies. This came to a halt in 2000 when a sudden downturn in demand exposed a major accounting scandal that caused the stock to tank and, after the inevitable inquisitions and firings, the company's assets were sold to several companies with the main telecoms interests being eventually merged with the French company Alcatel in 2006 to form, what was for a time, one of the current leading players in telecoms supply [6], although it too faded and its assets were eventually bought by Finland-based Nokia.

Many of the regional Bell operating companies, which became independent after the breakup of the Bell monopoly, were of a size comparable to a European country's telephony operator. In turn, they became active in diversifying their telephony business and, in several cases, became part of a larger grouping involving cable TV or mobile telephony. An example of this is Verizon, which was originally the mid-Atlantic states' Bell company, Bell Atlantic, and then went on to acquire NYNEX (the northeastern Regional Bell) and then GTE (an independent national telephone company) in 2000. Verizon also acquired the internet companies AOL and Yahoo and hence became one of the largest telecoms companies in the United States. However, it later changed its focus to mobility and sold off most of its former GTE landline business, AOL, and Yahoo, which were fading in importance.

Perhaps the most amazing footnote is that by 2007 much of the Bell network had reformed itself. Southwestern Bell Corp., later simply called SBC, acquired three of the other Baby Bells between 1998 and 2006, making it the largest landline player in the United States. It had also put together one of the largest mobile carriers, including one earlier spun out of a shrinking AT&T. In 2005, it acquired the remaining assets of the old AT&T Corp., mainly its long-distance network and internet business, and renamed itself AT&T Inc. Thus, something resembling the original AT&T company was recreated with 4 of 7 of its regional companies together with the long-distance network, making it once again the biggest telecoms company in the world. Two other Baby Bells had also merged into what had become Verizon, which also took over MCI, one of the few surviving significant long-distance communications companies, after the company which had bought it, WorldCom, collapsed in a scandal of accounting fraud. Thus, much of the United States was once again largely dependent on a company called AT&T as its monopoly telecoms supplier, but this AT&T was a rather different animal. Little of the public service ethos that permeated Ma Bell in its prebreakup days survived, and the 1996 Telecommunications Act greatly increased its freedom to operate. Growth and strategic takeovers made AT&T both the leading fixed-line and mobile phone operator in the United States [7]. As in the United Kingdom, despite the efforts of the central government, the monopoly phone operator proved to be too strong for any newly created competition. However, the landline was declining in importance and the critical U.S. mobile market later settled down to three major carriers, AT&T, Verizon, and T-Mobile U.S., with a number of small operators, fixed and mobile, mainly in smaller and rural markets. Cable companies also expanded into the local telephone market, although it was secondary to their internet and video distribution businesses. Government telecom competition policy seemed to settle down to a duopoly of internet/wireline providers, the traditional telecoms carrier and the cable company, with three or sometimes four mobile carriers. The rise in data communications was also largely driven by the U.S. technological developments as discussed in Chapter 4. Suppliers to this market were mainly new start-ups often emerging from California's Silicon Valley, which had become the hot spot for electronics and software innovation. Hence, suppliers such as Cisco were able to achieve worldwide presence. However, the data in Appendix A shows that the U.S. balance of trade-in telecoms equipment greatly deteriorated in the 50 years of this study. Despite the efforts of start-ups such as Cisco, U.S. telephony operators were importing significant amounts of equipment from suppliers such as Ericsson and Alcatel (later Nokia) while at the same time the many telecoms startup businesses of the late 1990s as well as the remnants of the former Western Electric were not able to achieve enough export business to compensate.

While this section is mainly about the United States, the position of Canada should not be overlooked. Given its closeness to the United States, it is not surprising that it largely followed the U.S. approach to telecoms rather than the colonial approach of many former members of the British Empire. The Bell Telephone Company of Canada Limited was established as a government-owned company in 1880 and had a license from the U.S. Bell Telephone Company and extensive rights to build telephony infrastructure in much of Canada. Northern Electric (later Northern Telecom, then Nortel Networks) was the main supply arm of Bell Canada, analogous to the role of Western Electric in the United States. The breakup of the U.S. Bell companies was not mirrored in Canada with the then-Bell Canada Enterprises Inc. (BCE) being largely analogous to a U.S. Baby Bell company. Since then, there have been several rounds of refinancing and restructuring of BCE. However, it continues to operate as the main telecoms business in parts of Canada, although Nortel was a casualty of a major financial scandal around 2008. There are currently two other mobile phone networks in Canada plus several regional telecoms suppliers making the situation similar to that in the United States.

5.2   Europe

In most European countries, from 1950 to around 1980, the telecoms network was run by a state monopoly usually with one or two major private companies as effectively monopoly suppliers to the operating organization. While this might be considered unhealthily close (given the ample scope for corruption in a relationship between a major supplier and a monopoly customer), it is a relatively simple arrangement to operate. In 1950, most major European countries had in place supply arrangements whereby its major telecoms needs were met by local suppliers if necessary, licensing the required technology. This was certainly true for France, Germany, Italy, Spain, the Netherlands, and Sweden. The dramatic technological changes and the need to produce mobile phone networks put a major strain on this concept. The changes for each country can be summarized thus:

France: The telecoms network in France up to the 1970s was widely accepted as a national joke. Bizarrely, responsibility for telephones fell to two different ministries, one for Paris and one for the rest of France. There were only 5 million lines in operation in 1974, half that in the United Kingdom, a country of around the same population of 53 million [8]. In 1975, a 5-year plan was adopted to modernize high-technology industry. In line with the overall "dirigiste" concept of French industrial development, there was extensive nationalization of the supply industry together with a large national investment targeted at achieving similar levels of telecoms availability to the United Kingdom and West Germany. Installing and operating telecoms equipment fell under the jurisdiction of the Direction General de Telecommunications (DGT), a state-owned arm of the PTT. Throughout this period, the major telecoms enterprise was Alcatel, which formed part of the vast Companie Générale d'Electricite (CGE) group, which had become nationalized in the 1980s, by which time over 30% of the country's manufacturing base was state-owned. As part of the nationalization push, foreign-owned suppliers such as ITT came under increasing pressure. While in the United Kingdom, the STC company was floated on the LSE with ITT progressively selling down its stake, in France, ITT's main company LMT was bought by the then independent French company Thomson in 1976. Thomson, in turn, was nationalized in 1981 and then merged with Alcatel. Thus, although in the 1950s much of the telecoms supply had been through subsidiaries of foreign companies such as Ericsson (Sweden) and

ITT (United States), by 1985, the newly merged Alcatel-Thomson company had 84% of the local market [8]. Not surprisingly, the extensively nationalized telecoms sector was inherently loss-making with its losses being covered by high telecoms fees and state subsidy. While these multiple changes of ownership took place, the country had to face the same issues seen in the United Kingdom for the choice of digital telephone exchanges. After a bewildering array of possibilities, not surprisingly the CIT-Alcatel E-10 system in 1977 had been adopted as the exchange of choice for the French network. However, as a result of the merger with Thomson and an even more complex merger with the telecoms interest of ITT in 1987, the CGE group emerged with three different digital exchange designs. Thomson's takeover of ITT included the well-regarded System 12 electronic exchange plus major business operations, notably the Belgium business known as Bell Telephone Works and the German SEL (Standard Electrik Lorenz) business. This gave the newly merged Alcatel-Thomson company around 42% of all European telephone switch sales.

This complex series of mergers and investment set what was now simply called Alcatel on the path to becoming a major worldwide supplier of telecoms intent on penetrating the U.S. market. With the later merger with Lucent from the United States described above, the French largely achieved their goal of having a major Tier 1 telecoms supplier but at a very large cost in terms of state investment and hidden subsidy from telecoms users. However, this position proved transitory with the Lucent-Alcatel business becoming part of Nokia in 2016.

In mobile telecoms, the initial development was under the nationalized telecoms industry France Telecoms. However, an independent second supplier was introduced. Initially, the French industry showed signs of following the United Kingdom's lead in developing a major world player (Vodafone) and a privatized France Telecom acquired Orange in the United Kingdom in 2000. However, the large funding needs for such a play may have given pause for thought and the U.K. business was subsequently merged with the U.K. operations of the German-led T-Mobile to create a new entity Everything Everywhere (EE), which, in turn, was bought by BT in 2015, which presumably had decided that divesting their U.K. Cellnet business to Telefonica was not such a good idea. Despite its retreat from the United Kingdom, France Telcom built on the Orange brand (Figure 5.2) and came to operate mobile and landline networks in several countries in Europe, Africa, and the Middle East, notably Spain, Poland, Egypt, and Morocco.

Germany: The history of German telecoms is linked strongly with Siemens, a large German industrial conglomerate dating back to 1847 with a worldwide presence in many areas of electronics and electrical engineering. By 1954, the national PTT was renamed the Deutsche Bundespost (DBP) with monopolistic powers over most post and communications. For the next 30 years, it operated

Figure 5.2 The widely used slogan (the future's bright, the future's Orange) that supported Orange's strong position in the United Kingdom and elsewhere. (*Source:* Orange.)

remarkably like the U.K. GPO with its own version of the Bulk Supply Agreement with fixed quotas for suppliers and obligation to cross-license technology. In this, there were two significant players and two smaller players. Siemens had 40% of the business, and the ITT subsidiary, Standard Electrik Lorenz (SEL) had 30% of the business. However, Siemens also had strong links to the two minor players DeTeWe and Telenorma. These cozy arrangements came under attack from the German monopoly commission and from other external suppliers in the 1980s, but these arrangements were in place while consumer demand in the 1970s greatly increased sales and the issues arising from digital switching seen in the United Kingdom and elsewhere. In developing electronic exchanges, the German industry suffered its very own "Highgate Woods" moment as late as 1979 when the internal development program was stopped. A crash recovery program led to the adoption of the Siemens EWSD system and ITT's System 12 being accepted in 1983 [9]. This meant that Siemens would struggle to maintain a strong position in the world telecoms industry while the acceptance of System 12 would open the door to Alcatel (when they took over the ITT business). Siemens did remain a major supplier of telecom carrier equipment into the early years of the new century, but put its telecoms interests into a joint venture with Nokia in 2007, and Nokia took full ownership in 2013.

Germany was late to privatize its PTT; the telecoms operation was privatized under the name of Deutsche Telekom in 1995 (with the German State retaining a minority interest). At this point, it was dominant in landline provision and its mobile phone operation under the name T-Mobile was the bigger operator in Germany.

While starting off with a similar position to France, it went in the opposite direction. The national Tier 1 supplier was the vast industrial conglomerate

Siemens. While it showed many signs of wanting to become a multinational player in telecoms, it was perhaps too widely focused to achieve this. Thus, it never seriously engaged in the U.K. market, despite being a joint acquiror of Plessey (with GEC) and steadily withdrew from other countries (such as Italy). Although it tried to carve itself a position in mobile phone handsets and electronic exchanges, these positions were fading by 2000.

Conversely, the T-Mobile mobile phone business (the operating name for Deutsche Telekom's mobile business) was successful in achieving presence in many major markets (including the United States), usually by buying out local suppliers. A competitor to the T-Mobile phone business was created by a second license being granted to Mannesmann (an established electronics group). This company performed well and was the subject of a bitterly fought takeover in 2000 by the U.K. Vodafone company. Thus, in 2000, the German mobile phone market had three strong international competitors in the shape of T-Mobile, Vodafone, and the Spanish Telefonica group (using the O2 brand).

Netherlands, Italy, and Spain: These countries found themselves below the scale necessary to support a Tier 1 telecoms supplier (without cross-border cooperation). Their position became like the United Kingdom where they started importing the key technology while seeking local content.

The Netherlands is home to the Philips company, one of Europe's main electronic conglomerates. In the 1950s to the 1980s, it was particularly strong in semiconductor development and consumer electronics. The Eindhoven Research Centre was regarded as a leading center in Europe for many developments. Inevitably, Philips developed telephone equipment, notably handsets and PBXs. However, it did not have a strong position in main digital telephone exchanges. In response to this, it formed a joint venture with AT&T to create APT (AT&T, Philips Telecommunications). Despite its illustrious parentage, this venture was not very successful beyond some orders from the Dutch. Target customers included the United Kingdom, which decided on the Ericsson AXE exchange as a second source to the United Kingdom's System X. AT&T started looking at joint ventures elsewhere (such as Italy) and eventually bought Philips out of APT in 1990. With that, Philips ceased to be a major player in telecoms and had largely withdrawn from consumer products. Its main concentration is now medical products.

Meanwhile, the Dutch telephone system had started to acquire Ericsson's AXE exchanges.

In Italy, not surprisingly, the telephone network operating arrangements are extremely complex but effective. On the supply side, local manufacturing is mainly provided by local subsidiaries and affiliates of Siemens, Ericsson, and Alcatel. The main supplier was Italtel, initially a Siemens operation but later to join forces with AT&T to offer their digital telephone exchange equipment. Little has been seen of exports of telecom equipment or services.

Spanish equipment supply has historically been mainly locally produced versions of international equipment from ITT, Ericsson, and Alcatel. However, Spain has produced a major international supplier of telephone services. Telefonica started business in 1924 when the Spanish government granted a monopoly concession to operate their entire telephone network to Compania Telefonica National de Espana SA (CTNE), which belonged to the newly formed U.S. ITT company (see Section 5.4.3). The company was effectively nationalized by 1945 and subsequently partially divested by the Spanish state so that in 1986 it was renamed a much more manageable Telefonica, in which the Spanish state still held about 36% of the equity [10]; the renamed business took stakes in a significant number of telephone operators, notably in Chile, Mexico, and Argentina, as well as acquiring the Cellnet mobile phone business from BT, which is still a major U.K. network operator under the brand O2. Not surprisingly, Telefonica has now become a large part of the Spanish industrial state with all the political and economic issues that go with that position.

Nordic countries: This is possibly the most interesting area for review. Few observers of world telecoms in 1950 would have predicted that two of the most significant companies in world telecoms in the next 50 years would emerge from Nordic countries (Ericsson and Nokia). In many ways, the very fragmentation of the area with five relatively small countries (Denmark, Norway, Iceland, Sweden, and Finland) makes an international perspective imperative for any technology company to succeed. Remarkably, even within individual countries, the national PTT did not enjoy a clear legal monopoly in Sweden and Finland, the parent countries of L. M. Ericsson and Nokia, respectively. The early cooperation among the Nordic countries in the development of international mobile phone networks was one factor that led to the rapid growth of the two champions.

However, certainly in the case of Ericsson, the seeds of success were set much earlier with a strong technological base and a pragmatic approach to international sales where it would often penetrate a new market (such as Australia) by offering to be a reliable second source supplier, frequently opening local subsidiaries to strengthen its presence (in 2000, Ericsson reported that fewer than 40% of its employees and 3% of its end-user sales were in Sweden) [11]. Remarkably, for most of the 50 years reviewed, Ericsson was the second source supplier in Sweden, the local PTT, Televert, providing more equipment from its in-house manufacturing. Even before the development of digital exchanges, Ericsson was able to offer Crossbar switches, which performed better than the Strowger switches provided by U.K. companies. The introduction of the AXE digital electronic exchange (again, surprisingly developed in conjunction with Televert) strengthened Ericsson's position and it achieved sales in many countries including Australia, the Netherlands, Mexico, Brazil, the United Kingdom, and the United States. Ericsson's strong position in radio technology put it in a leading position to supply equipment to new mobile phone networks (by 2000, 73% of its sales were for mobile systems).

The story of Nokia is again surprising. It arose from a Finnish conglomerate whose main business was wood products. A diversification plan led them to start supplying telephone equipment to a local Finnish network. While its switching equipment has been sold throughout the world, Nokia was primarily known for its mobile phone handsets. Due to the wide acceptance of GSM phone protocols, Nokia was able to achieve world leadership in mobile phone supply in the period around 2000 with a supply of over 100 million handsets a year. However, its position with handsets was eroded as it failed to maintain its lead with the introduction of smartphones and its phone business was eventually bought by Microsoft (to promote their mobile operating system) in 2014. This enabled Nokia to concentrate on its equipment business, which went on to merge with the Lucent-Alcatel business in 2016. By 2017, the rebranded Nokia network business was the third biggest telecoms equipment supplier in the world (after Huawei of China and Cisco of the United States).

5.3   Japan

The history of modern Japanese telecoms effectively starts in 1952 when the telephone system that had been run by AT&T after World War II was transferred back to a state monopoly organization Nippon Telegraph and Telephone (NTT). This steadily rebuilt the telephone infrastructure using local suppliers. In the usual Japanese manner, supply was through a group of indigenous companies, with the main players being NEC, Matsushita (now called Panasonic), Oki, Hitachi, and Fujitsu. Nippon Electric Company (NEC) was the lead supplier. NEC was and is an enormous conglomerate with strength in computing and semiconductors. Its history in telecoms goes back to the 1904 and, before World War II, it had largely modeled itself on the Western Electric arm of AT&T and produced Strowger-like switching equipment. As the main supplier to NTT, NEC was able to successfully develop its own Crossbar switching system in the 1950s and then embarked on a joint development of digital switching being led by NTT. The development of the DEX series of electronic exchanges commenced in 1964 and was undertaken by NEC, Matsushita, Oki, Hitachi, and Fujitsu, with the lead from NTT. It is a tribute to the effectiveness of intercompany cooperation in Japan that a development of this complexity involving six supposedly independent companies was carried through with the installation of fully digital electronic exchanges starting in the 1980s. Despite the apparent technical success of this development, relatively few sales were made outside of Japan, perhaps for the same reasons that affected the U.K.

System X sales, including the lack of one clear commercial leader to manage the complex sales process.

As has been previously described, the development of mobile telephony in Japan was highly fragmented with several competing incompatible specifications. Thus, although the Japanese industry was clearly capable of achieving a dominant position in mobile phone supply, most chose to concentrate on the specific needs of the local market. Hence, European makers were able to take a lead in many territories. Highly innovative mobile data applications under the i-mode brand were developed in Japan and tailored to the specific market needs of the Japanese consumer [12] (see Figure 5.3). The need to facilitate Japanese language characters and the local commitment to emojis meant that early phone development was based on the local Itron operating system, with makers being late in adopting internationally successful operating systems such as Android. As global technology became more accepted in Japan, locally developed features were overtaken by the TCP/IP-based system adopted in Europe and America.

The main mobile telephone operator in Japan is DoCoMo, in which NTT had a majority interest after it floated DoCoMo on the Tokyo Stock Exchange in 1998 (NTT was itself floated in 1987 with the Japanese state retaining around a 30% stake). However, it experienced a weakening position with more aggressive new suppliers gradually growing market share. An example is

Figure 5.3   Flip phones were popular in Asia. (Source: Scian Cho, Shutterstock.)

Softbank Mobile, which purchased Vodafone Japan in 2006 and was the first to offer Apple iPhones to the Japanese market. NTT responded in a way curiously reminiscent of BT's reentry into the mobile phone market in the United Kingdom by taking over all of DoCoMo in 2020 [13].

Thus, despite Japan possessing a strong technological base and the largely protected home market in which to evolve new product innovations, it did not build up a strong position as a world telecoms supplier beyond the supply of peripheral items such as handsets, facsimile machines, and PBXs. Even in this area, its position is under threat from lower-cost producers such as Korea and China.

5.4   The United Kingdom

While one might expect the United Kingdom to have been an early adopter of telecoms, in practice, this was not really the case in the period from the invention of the telephone (1876) to the start of my main period of interest (1950). Some of the factors causing this have been discussed in previous chapters (notably Section 2.2). Once the U.K. Post Office had achieved an effective monopoly on telecommunications in the United Kingdom (1911), it did little to stimulate market growth with relatively high call charges and poor service.

A feature of the U.K. situation was the relationship of the supply industry to the telephone operator (GPO). In the United Kingdom, there was a slightly different mindset that was manifested in the Bulk Supply Agreement that applied to the industry between 1933 and 1969 (see Footnote 1 in Chapter 2). Implicit to this arrangement is that the design authority for the network would be the GPO with suppliers operating simply as cost-plus subcontractors. This approach led to a reluctance on the part of suppliers to innovate; indeed, a surprisingly large part of the early technology used in the telecoms network came from licensing non-U.K. technology, particularly from the United States. Thus, the supply base of the U.K. telecoms industry in 1950 was based around the five original members of the Bulk Supply Agreement:

1. *Siemens Brothers, Woolwich:* Originally set up by the German Siemens family, but by 1950 it was an independent company in which the U.K. conglomerate AEI had a stake;

2. *British Ericsson, Nottingham:* Originally a spin-from the Swedish telecoms company;

3. *Automatic Telephone Manufacturing (ATM), Liverpool:* Originally a licensee of U.S. telecoms equipment;

4. *Signal Telephone and Cables (STC), Southgate:* A subsidiary of the U.S. International Telephone and Telegraph (ITT) company;

5. *GEC, Coventry:* Possibly the only company that did not clearly trace its routes to a non-U.K. parent.

These companies were destined to form the core of the U.K. telecoms supply industry in the period from 1950 to 2000. All five were heavily imbued with the mindset of the BSA. The work was shared out evenly between them with two sites (Southgate and Liverpool) being used to establish the cost basis for any orders. Two companies (Ericsson and ATM) were independent and probably subscale to undertake any serious product development. The balance was part of larger diverse conglomerates who were not primarily focused on the U.K. telecoms industry.

In this chapter, we will explore how these five organizations emerged as the three key players in U.K. telecoms supply by the 1980s: GEC, Plessey, and STC. We will also look at two other companies who had a significant role in the period: Racal and Pye TMC.

As the five most significant U.K. telecom suppliers in the period, I show the turnover and telecoms sales for them over the 50 years in Appendix C. From this, it can be seen that these suppliers were either small (when compared with overseas competitors) or that telecoms was only part of their revenue base. Below, I expand on the history of these five key suppliers to the U.K. industry.

5.4.1 GEC

The history of GEC and its remarkable CEO, Arnold Weinstock, has been the subject of numerous fascinating books, notably by Aris [14] and Crowe [15]. In many ways, it mirrors Britain's industrial policy of the postwar years. The 1960s saw a dramatic saga of mergers and rationalizations followed by a period of incredible financial strength. A much-discussed cash balance of £1 billion in the 1980s led to it being regarded as Britain's most important industrial firm. However, this strength was not successfully exploited, and the company came to an end with a humiliating decline and near-total disappearance by 2004.

The history of GEC went back to the 1900s when it was created by its founder Lord Hirst from an assembly of electrical businesses initially as a wholesaler and distributor of electrical goods [14]. In the 1920s and 1930s, it became one of the United Kingdom's preeminent electrical manufacturers, with products including that of telecoms equipment supply based in Coventry. Since Hirst's death in 1943, the business had stagnated, so its history really restarted in 1961 when Arnold Weinstock joined the company. This came about when GEC acquired a much smaller TV company, called Radio and Allied, which Weinstock was running on behalf of his father-in-law, Michael Sobell. Weinstock himself came from a typical North London Jewish background where he trained as an accountant before his father-in-law persuaded him to run what was to become Radio and Allied. Weinstock clearly impressed all around him and made the company successful in what was a difficult market. To a considerable extent, GEC was attracted to the Radio and Allied acquisition to acquire Weinstock as much as the company. They paid a considerable premium for a much smaller company, and Weinstock became a senior player in GEC (he and his family held 14% of the post-merger GEC) initially in a joint CEO role, but he soon assumed absolute power [16].

GEC was, by 1961, a struggling company with a wide range of products and a bureaucratic structure that stifled most opportunities for success. In the first 2 years of his regime, Weinstock drastically rationalized and focused the company. Together with a small group of henchmen, he shut the pretentious London headquarters and greatly simplified the organization so that line managers were very clearly aware of their responsibilities and rapidly came to know that their performance would be scrutinized thoroughly and regularly. This had the effect of dramatically improving the performance of the company.

GEC's big move came in 1967 when it made an audacious bid for Associated Electrical Industries (AEI), a much bigger conglomerate operating broadly across the U.K. electric industry. Both companies were quoted on the London Stock Exchange and what followed was a classic battle between the established company with a rather poor record and a smaller competitor that could point to the improved performance achieved in GEC in the previous years. In a closely fought takeover battle, GEC prevailed and took over the whole business and set about imposing much more rigor to reviewing its operations than it had previously seen [17].

The telecoms sector was only one of the areas of the electrical industry that the new conglomerate covered (other notable sectors included power generation, TV, lighting, and cables), but it became very quickly apparent that a dramatic improvement in productivity in this sector was both necessary and achievable. It did not take long for this to manifest itself in a plan to shut down the AEI Siemens Brothers Works factory at Woolwich in the eastern suburbs of London (Section 2.4). If the battle to take over AEI was a milestone in the history of stock-exchange takeovers, then the closure of Woolwich was similarly a milestone in the U.K. industrial history, a site of 40 acres, which, at its peak, employed 75,000 people was shut (Figure 5.4). Despite strong opposition from unions and politicians, Weinstock and GEC prevailed possibly because their logic was extremely clear and they refused to be bullied. While a recently elected Labour government might have been expected to intervene, Weinstock always enjoyed a good relationship with Labour politicians. They also made reasonable redundancy arrangements and acted professionally throughout. So Woolwich was shut in 1967 and the bulk of the telecoms activity moved to the Coventry site of GEC. Remarkably, the merged GEC company retained its 40% share

Figure 5.4   Siemens, Woolwich, part of Britain's (fading) industrial heritage. (Source: Kleon3 CC-SA 4.0.)

of the Bulk Supply Agreement as they had notionally two plants within their organization.

A year later GEC was involved in another significant takeover. However, the acquisition of English Electric went through without a major battle. The English Electric management accepted the need for more professional direction and GEC was increasingly seen as a major player in the United Kingdom's industrial policy. While the power generation business of English Electric was the trigger for the move, GEC acquired some of the best electronics technology available in the country with English Electric subsidiaries such as Marconi and Elliott Automation. The GEC merger also facilitated the U.K. government's plans for rationalizing the U.K. aerospace and computer industry sectors, both of which were strongly represented within the English Electric organization and were hived off into newly formed groups.

Thus, by 1970, GEC was an extremely strong industrial conglomerate with interests in power generation, telecoms, defense, and consumer products. However, its very broad scope meant that, despite its large size, it was still subscale in many areas (including telecoms). In several areas, GEC approached these issues by a range of strategic partnerships and divestments that are outside the scope of this work. However, the culture of GEC with its extreme focus on short-term operating performance could discourage strategic thinking, and this was perhaps most notable in the relatively weak engagement in the semiconductor industry, which was crucial to the onward development of telecoms (and defense).

The 1970s was a crucial time for the U.K. telecoms industry. It was clear that there was an urgent need to develop a new generation of electronic telephone exchange, and the issues relating to System X have been previously described in Section 2.3. Timing was not good for GEC; it had a wide range of issues to resolve as it digested and rationalized the various parts of AEI and English Electric into its own organization. The financial performance of GEC remained strong and hence there was no compelling need to engage in further mergers. While various players (including the TUC) [18] had advocated for a merger of all the country's telecoms manufacturers, GEC was too broadly spread to take the lead in this, but at the same time too financially and politically strong to agree to a minor role in such an entity.

It was not until the mid-1980s that GEC came up with what was an inevitable suggestion to rationalize U.K. telecoms; namely, it would take over Plessey, its archrival [19]. By then, most of the damage had been done in terms of the delays in introducing System X to the world market and the failure to establish an effective export marketing activity for U.K. telecoms. Even this proposal was blocked by the U.K. government on the grounds that it would give GEC a near monopoly in the supply of defense electronics. Ironically, about 15 years later, British Aerospace achieved this very situation with the U.K. government by acquiring GEC's then-defense electronics business [20].

In the 1980s, the development of mobile phones largely passed GEC by. The culture and organization of GEC would discourage serious innovation and start-up businesses. (This was described to me by Peter Rowley, a former colleague of Weinstock, at an interview on August 15, 2022.) It was left to Racal, which had considerably less financial and technical resources, to successfully innovate and eventually spawn a world-leading player in this area, Vodafone.

In 1989, the inevitable happened; as development work for System X began to run down, Weinstock could finally carry through the acquisition of Plessey. However, to make this acceptable, he had to operate in collaboration with the German Siemens company. The telecoms interests of GEC and Plessey had already been merged into GPT (GEC, Plessey, Telecoms) [19], a 50:50 company. However, GEC eventually achieved a full takeover of Plessey in 1989 with Siemens taking a stake in GPT. Despite having a strong technical base of its own, Siemens did not seem to engage seriously in this arrangement and there was little flow of products either way between the partners. Nor was the merged company of adequate scale, it rated around ninth in the world telecoms industry with 3% of the market [21].

For GEC, the late 1990s had one key question: Who would succeed Weinstock? Initially, it was felt that a dynasty would emerge with his son Simon being groomed for a senior position. However, a medical issue intervened and eventually George Simpson was appointed in 1996 as CEO to succeed Weinstock, who retired to look after his racehorses. Simpson (formerly with Lucas and a previous deputy CEO of BAE) seemed to be ideally suited to take on this role of leading a flagship company of the U.K. economy [22]. His arrival coincided with a strong market sentiment that GEC should rationalize its very diverse holdings and deploy its legendary "cash mountain" to better effect. After some initial rationalization, the initial thought was to divest its GPT telecoms business to its joint owner, Siemens, with the eventual aim to acquire BAE and become a major world defense supplier. However, Siemens balked at the deal, so in a significant U-turn, GEC decided instead to concentrate on its telecoms business (as explained by Rowley during his interview). Few people expected that within 7 years GEC would have been all but destroyed. In a series of bold moves, GEC:

- Sold its defense electronics business (largely trading as Marconi) to British Aerospace for $7 billion;
- Liquidated its joint interest in its power generating business with its French partner Alsthom;
- Sold off its consumer electrical interests;
- Bought out its telecoms partner Siemens from its GPT business;
- Changed its business name to Marconi;
- Made two major acquisitions of telecoms systems companies in the United States: Fore and RelTec.

To say these changes did not work out is an understatement. With hindsight, it seems that the sell-off prices were too low and the acquisition prices were too high. Even at the time, it must have been clear that the reequipment of the U.K. network with System X had pretty well finished and there were few export orders in prospect. By 2001, it was clear that the strategy was failing, with the results from the U.S. acquisitions disappointing [23]. A sad footnote to developments at Marconi was a comment from Weinstock to Rowley at the end of 2001: "It isn't like when we had cash in the bank, Mr. Rowley." Weinstock died in July 2002.

The final straw was when in 2004 BT put out a tender for exchange equipment for which Marconi had bid, but it was won by Huawei, the up-and-coming Chinese competitor. Shortly afterwards, Marconi was restructured giving around an £8 billion destruction of shareholder value in 7 years. What should have been the United Kingdom's Tier 1 telecoms supplier ceased to exist apart from pieces that were mopped up by competitors and a rump organization that became a private company named Telent.

The significance of GEC as a standard bearer (and repository) of the U.K. electric industry is illustrated by the family tree in Figure 5.5 which shows how companies came together to form the GEC group of the late 1990s.

5.4.2 Plessey

While one person dominated GEC, Plessey was the base of the Clark family dynasty. The beginnings of the company go back to 1917, with the Clark family emerging as dominant shareholders in the 1920s [24]. Initially, the company made radio components and was a small-scale supplier to the GPO of handsets and similar equipment mainly coming from its base that was established in Ilford, Essex, in 1921. However, the company gained a reputation for technical excellence and thrived on government contracts. During and immediately after World War II, the company repurposed a whole string of stately homes that had become research establishments into a curious combination of research facilities and factories. Sites included:

- *Caswell:* Northamptonshire (its first acquisition in 1940);
- *Roke Manor:* Romsey;
- *West Leigh:* Hampshire;
- *Taplow:* Buckinghamshire (Figure 5.6).

The leadership of Plessey spanned three generations. Byron Clark (born in 1875 in the United States) was involved in the founding of the company (named after the Northumberland birthplace of the wife of a cofounder). His son Allan Clark (born 1902) took over the leadership in the 1930s before passing the role over to his oldest son John (born 1926) in the 1950s, who had his brother Michael as sales director. To describe the management style of the Clark family as autocratic would be rather an understatement, particularly the patriarch Alan Clark [17]. Even by the standards of the time, his political views and management style were regarded as extreme. However, with this style also went considerable entrepreneurial flair and a drive to acquire modern technology. What was a key factor in how the history of the U.K. telecoms industry played out was the antagonism felt between Plessey led by the Clark dynasty and GEC ruled over by Weinstock. This went beyond the inevitable competition between the two major suppliers to U.K. telecoms (and defense).

Up to the 1960s, the company was a small player even by the standards of the U.K. electronics industry. Sales in 1935 were around £800,000 and by 1950 had only reached around £5 million, the business including domestic radios, electronic components, and some basic telephony equipment such as handsets. It was thus already a member of Telephone Equipment Manufacturers

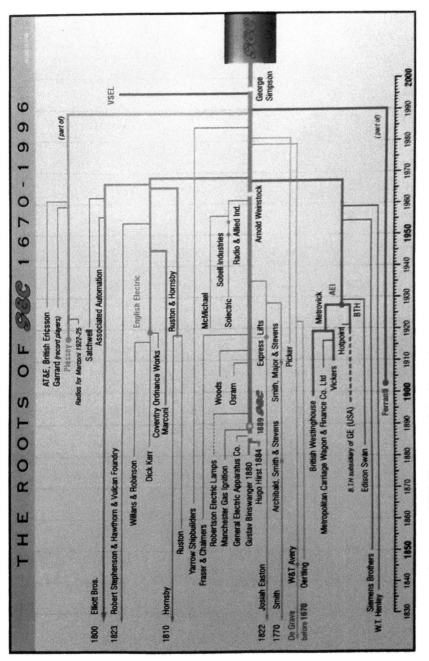

THE ROOTS OF *GEC* 1670-1996

Figure 5.5 The illustrious family tree of GEC. (Source: Bodleian Library.)

Figure 5.6 Taplow, a magnificent Victorian pile used by Plessey up to 1987. (Source: Troxx CC 2.5.)

Association (TEMA) and a party to the GPO Bulk Supply Agreement. However, in 1961, the company moved to become a major player in U.K. telecoms when they bought out the two independent businesses supplying exchange equipment within the GPO Bulk Supply Agreement; British Ericsson at Beeston, Nottingham; and Automatic Telephone & Equipment (AT&E) at Liverpool. With this, Plessey become a 40% supplier under the BSA.

Throughout the 1960s, the company was keen to grow through acquisition and developed a strong position in the U.K. defense industry taking on key projects in radar and sonar. It also bought Garrard, a major supplier of record-players. Of more interest were its forays into the semiconductor industry. In line with its autocratic management style, the Clarks initially did a license deal with Philco without much consultation with their managers for surface barrier semiconductors that eventually proved unsuccessful [25]. However, the company then entered several other license deals and established what was to become possibly the United Kingdom's most successful semiconductor facility at Swindon in Wiltshire with the first planar integrated circuits being produced in 1965. With support from several military contracts, Plessey had the technology for both analog and digital silicon integrated circuits to serve its activities in defense and telecoms. In comparison, GEC virtually exited semiconductor manufacturing in 1969.

Throughout the period, Plessey maintained a strong involvement in the semiconductor industry and received significant U.K. government support for its developments. Thus, when competitors such as Ferranti encountered problems, Plessey was encouraged to take over their activities.

The semiconductor interests of Ferranti were acquired in 1988, strengthening Plessey's position as the United Kingdom's leading semiconductor manufacturer.

The company remained fragmented over many locations spread throughout the United Kingdom and did little to rationalize this structure. In its telecoms businesses, the Beeston plant concentrated more on transmission and the Liverpool plant concentrated more on switching. Unlike other U.K. telecom suppliers, Plessey also had several subsidiaries in export markets, such as Portugal and South Africa. These began to give the feedback that the indigenous U.K. switching product was no longer attractive to export customers. Plessey thus decided to support development work on Crossbar switching, which had been initiated at its Liverpool factory, despite the active disinterest of the GPO. In 1967, this work bore fruit with first orders for their 5050-Crossbar system coming from the GPO. However, to smooth the path, Plessey had to license its product to its archrival GEC.

When the GPO finally commissioned the development of System X, Plessey was the preferred contractor. However, they did not have the scale or resources to do the work without GEC. Their fragmented structure meant that, even within Plessey, the work was split between around 10 different locations (including Liverpool, Poole, Taplow, Caswell, Roke Manor, Beeston, and Ilford), which hardly made for a simple project management situation.

Although most of its telecoms revenue was in the United Kingdom, Plessey retained ambitions to attack the U.S. market, and in 1982 bought a long-established U.S. supplier of office switching equipment, Stromberg-Carlson. This acquisition was not particularly successful, and when the cash flow from System X began to slow, its disappointing results were a further drain on Plessey's credibility with shareholders. Stromberg-Carlson was sold to Siemens in 1991 as part of the post-GEC acquisition arrangements.

Despite growing steadily from the 1950s to a total revenue of £208 million in 1970, Plessey was always subscale for a major project like System X. When the newly privatized BT started to order some of its requirements from Ericsson rather than just buy System X, it was clear that Plessey would not be sufficiently strong to become a Tier 1 telecoms supplier and their purchase by GEC, although protraction was inevitable [26]. GEC's first attempt to take over Plessey was vetoed by the U.K. government in 1986 primarily because of concerns over creating a monopoly on defense contracts [20]. Instead, Plessey and GEC finally reached an agreement to merge all their telecoms interests to form GPT (GEC Plessey Telecoms) in 1988 [27]. The poor relationship between the Clarks and Weinstock at GEC made negotiating the deal difficult. A masterstroke from Weinstock was to agree that John Clark could chair GPT, but, in exchange for this, both parties agreed that they would have first right of refusal if the ultimate ownership of the other party's stake in GPT changed. In practice, this made Plessey unsalable since a large part of its revenue was now tied up in the GPT deal. The 50:50 power split between the Clarks and Weinstock was never going to be comfortable and eventually GEC prevailed with a full takeover of Plessey in league with Siemens in 1989. The deal valued Plessey at around £2 billion [28].

5.4.3   STC

The roots of STC can be traced back to the start of telecoms with the invention of the telephone by Bell. STC emerged from the early days of telecoms to be acquired by the Bell company of the United States but was then part of the extraordinary buyout of ITT by Sosthenes Behn, which occurred in 1925 [29] when Behn was able to raise $29 million from J. P. Morgan and Company to buy out virtually all the oversea interests of AT&T. This deal was agreed by AT&T to stave off the continuing threat of a U.S. antitrust action.

By most standards, this was a remarkable deal for ITT. For its investment, it acquired major telecoms suppliers in a wide range of territories including Germany (SEL – Standard Electrik Lorenz), France (LMT), Belgium (Bell Telephone), and the U.K. STC company. Throughout the 1920s to the 1950s, these entities operated with considerable freedom in their own territories (very useful if they are on different sides of a war). This stance seemed to work well with most European countries (including the United Kingdom) usually regarding the local ITT subsidiary as part of the indigenous supply industry.

STC developed a strong position in the production of cables for telecoms and other applications. Work in this area dated back to before the invention of the telephone and the United Kingdom had an active interest in rolling out undersea cables to provide communications within its vast empire. What is rather surprising is that STC also became a founder supplier to the GPO under the Bulk Supply Agreement. This was despite the GPO taking the conscious decision in 1922 to not use the AT&T technology for switching but to concentrate on Strowger-based switching that was licensed by several U.K. suppliers such as ATM in Liverpool. STC set up a telecoms plant in Southgate, North London, which became a key part of the Bulk Supply Agreement.

However, by the end of the 1960s, STC were at a considerable disadvantage as a major telecoms supplier in the United Kingdom. Both GEC and Plessey had secured 40% of the BSA by their acquisitions while STC retained its 20% share. Perhaps a bigger issue was STC's ambiguous position concerning ITT products. When it came to development, new technology exchanges the GEC and Plessey regarded STC with suspicion probably with some cause. STC could offer the ITT Series 12 electronic exchange as a possible alternative to the locally conceived System X. While GEC and Plessey had their suspicions of each other they were even more wary of STC gaining access to their developments.

The early development of System X involved STC as well as Plessey and GEC. However, it was widely accepted that STC's efforts to develop a key transmission interface were not successful, and the issue of how to go forward with the development of System X was debated throughout the United Kingdom up to the then-Prime Minister Jim Callaghan who in 1978 agreed in principle that Plessey should take over STC. The general perception was that Plessey had good technology but poor management while STC had strong management but poor technology. In the political turmoil of 1978, this decision disappeared and it was left to the GPO to set up an alternative arrangement that would achieve the phased withdrawal of STC as a front-line telecoms supplier.

Meanwhile, the ITT company had begun to explore how it could develop its structure and generate value for shareholders. This led to it placing minority stakes of its country subsidiaries (including STC) on local stock markets. Hence, STC was floated on the LSE in 1979, with ITT holding steadily reducing to around 24% of the equity which was then bought out by Northern Telecom (originally a Canadian offspring of the Bell family) in 1987.

The deal struck by the GPO with STC suited ITTs divestment program very well. As a "golden goodbye" from System X, STC gained most of the rights to manufacture the interim reed relay exchanges TXE4 and TXE4A. At the same time, they ceased to be part of the System X development. The inevitable delays in System X created a double bonus for STC; not only did they not have to face the costs and disruption of System X development, but the life of their TXE4 products was also dramatically improved as timescales for System X stretched out. This, in turn, enhanced the short-term profitability of STC and hence its share price.

The burgeoning share price not only allowed ITT to unload its holding but also enabled STC to acquire one of the United Kingdom's key technology players (ICL) for £430 million in 1984. ICL was the U.K. standard-bearer in the world computer industry being the result of many mergers of early U.K. computer businesses. The rationale for STC acquiring the business was, to say the least, questionable. The CEO, Ken Corfield, apparently acted without a lot of input from other company managers based on the expected convergence of computing with telecommunications. While this convergence was likely, the general impression within the company was "this is the wrong end of the computer industry." ICL was primarily involved in large information technology (IT) systems for U.K. government and big corporate clients.

The inevitable conclusion was that little synergy was found and, within 5 years, the bulk of the profits of the merged group was coming from the old ICL business (the sales of TXE exchanges were almost finished by 1989). With this, came an inevitable pressure to demerge. Eventually, deals were struck to take the ICL business into Fujitsu (ICL's Japanese partner) while the residual telecoms business was eventually acquired in 1990 by Nortel, as the Canadian company Northern Telecom had been renamed [30]. Other parts of STC were sold, with the world-leading submarine cable business going to Alcatel in 1994 for £600 million.

Nortel itself did not survive long into the twenty-first century. It was caught up in a very convoluted series of self-destructive acts involving insolvency, accounting scandals, and pension liabilities. The North American market for traditional telephone gear had also slowed down, and Nortel's market share in the growing internet gear market was far behind that of leader Cisco. This led to a series of sell-offs of Nortel's assets, the biggest being its patent portfolio. By 2010, most of the assets had gone, but litigation about pension liabilities (including those owed to STC employees) continued into the following decade. STC Southgate, the flagship site of the STC telecoms business, is now a housing estate.

5.4.4 Pye TMC

The Telephone Manufacturing Company (TMC) had a long history of supplying the U.K. telecoms industry. As the name suggests, TMC was primarily involved with the manufacturing of telephones. Like many U.K. telecom companies, their roots lay in importing phones and other telecoms equipment for the U.K. market, in this case from Germany. World War I forced them into making their own products, and in 1920 the business was floated on the U.K. Stock Exchange as TMC. At the time, the main activity was making phones in the London borough of Dulwich, although they had widely spread operations including one in Australia. Later phone manufacturing was moved primarily to Airdrie in Scotland with Dulwich remaining the engineering base.

The business continued supplying telephone and small private telephone exchange equipment (PBX) to the U.K. market. Some were supplied to the GPO, which held a monopoly on the supply of telephone and small PBXs (under 100 lines) up until the 1980s. In 1929, the company had split into a sister company, Telephone Rentals, which again did much as its name suggests renting telephones and small PBXs to the U.K. market. Telephone Rentals remained TMC's major customer. The company never had enough resource or scale to be a major phone supplier but became very adept at developing specialized equipment for niche telecoms markets (such as railways). Another example of its engineering expertise and relationship with the market was that it supplied the "Speaking Clock" to both the U.K. and Australian markets [31].

As a supplier of telephones to the GPO, TMC had good access to the U.K. market and clearly established a good technical rapport with the GPO. As part of this, it proved capable of developing equipment to enable the GPO network of Strowger telephone exchanges to be upgraded as customer demands increased from the 1960s. However, the 1960s was one of business consolidation. Telephone Rentals, TMC's major customer, began importing PBXs from Ericsson in Sweden. In 1960, TMC was the subject of a fiercely fought takeover battle, which ended up with it being taken over by Pye, a larger electronics group that itself had a long history of supplying radio equipment. What was remarkable was that the losing consortium in this battle was a company created by the seven other companies that were then party to the GPO Bulk Supply Agreement. This blatant attempt to stop an outsider company entering the special relationship of the BSA caused considerable comment in government and elsewhere, all for a business that at the time had a turnover less than £3 million [32]. The battle was lost on price and the consortium went on to buy a much smaller company (Phoenix Telephone) as a sort of consolation prize.

However, the Pye group began to struggle in the 1960s as a result of competition in the supply of domestic televisions a major part of the group. In its search for an exit from this business, it approached Philips, the Dutch multinational electronics conglomerate, who was already a supplier to it. This led to Philips acquiring Pye in 1967, apparently on the grounds that its product range would fit within the Philips products.

TMC moved its headquarters to Malmesbury in Wiltshire in the 1970s and closed its Dulwich site. Meanwhile, most manufacturing was relocated to Airdrie in Scotland. With the phasing out of the Strowger exchange equipment that started in the 1970s, TMC concentrated more on PBX development and supply. It successfully developed and marketed the Ambassador and later the Herald PBX, which was successfully sold in the United Kingdom by BT, which at the time had a sales monopoly for PBXs below 100 lines. The manufacturing of these products was done in expanded sites in Scotland [33].

Given that it was a subsidiary of a major telecoms multinational, it might have been expected that TMC would become a major supplier of this technology into the U.K. market. However, Philips had its own plans and agenda that did not seem to involve the U.K. telecoms market. Its PBX supply was very fragmented, and it was clear that, with liberalization, the market would become much more competitive with suppliers such as Mitel of Canada entering the market. As part of its attempts to rationalize its telecoms activities, Philips struck a deal with the U.S. AT&T group to market its products to the Dutch PTT and then the United Kingdom. The TMC Malmesbury site was thus repurposed to software development for what later became Lucent and the residual U.K. business was banished to Scottish sites.

The remaining Scottish activities steadily wound down as they were very much on the periphery of the Philips/Lucent vision and was eventually shut down with residual activities bought by Mitel of Canada from their base in Newport, Wales, in 1995. Mitel later became a victim of the post-2000 crash, but some of it survives to this day. Meanwhile, the Malmesbury engineering site was closed in the early 2000s and is now a housing estate. Lucent later became part of the French Alcatel group in 2006. Ironically, a small operation under the TMC name survives to this day in Australia.

The Pye TMC saga gives a snapshot of how the U.K. supply industry was affected by the rapid liberalization and globalization of telecoms that started in the 1970s. The TMC business and its clear capabilities had served its purpose and without the support and investment needed for future products its lack of future was inevitable.

5.4.5 Racal

The preceding four companies were all longtime, established businesses whose roots were in telecoms in the early twentieth century (and in the case of STC even earlier). They all disappeared around 2000 and left very little trace of their input to the United Kingdom's telecoms industry. Racal is different. It started much later (1950), and although the Racal company also disappeared around 2000, it left a legacy in the shape of Vodafone, still one of the world's major mobile phone operators. It is tempting to argue that its late start-up and remoteness from the history of the U.K. telecoms industry were key factors in generating its success.

The company was founded as a consultancy in radio communications. From its early days in west London, it moved to what was to become its base in Bracknell, Berkshire, in 1954. It won early contracts for defense radio and after being listed on the LSE in 1961, also moved into TV transmission. Like other key players in telecoms, its history was dominated by one leader, Ernie Harrison, who became chairman in 1966. He served as chairman throughout its key period when the company expanded rapidly and made the key decisions that led to the creation of Vodafone.

As a well-regarded technology company, Racal expanded rapidly with several acquisitions spread through the U.K. electronics industry [34]. These included:

- *Airmec (1969):* A controls company;
- *Amplivox (1971):* A maker of hearing aids;
- *Decca (1980):* A maker of navigation systems, involving a tough take-over battle with GEC;
- *Chubb (1984):* A security company.

However, the key actions as far as telecoms were concerned related to its involvement with data communications over the telecom network. It started by marketing the U.S. Milgo modems in 1968 and later formed Racal-Milgo to manufacture the products. It also set up Racal Mobilcal, which was involved in mobile data communications.

The key move in telecoms was made in 1983 when Racal won one of the two initial licenses for mobile telephony in the United Kingdom. It created Racal Telecom, as a subsidiary of which it held an 80% holding to set up this business with Millicom Inc. (a U.S.-based communications group) as a partner to improve its capabilities in this new area. Racal Telecom successfully operated this business and won successive licenses for later generations of mobile phone networks. To raise capital, 20% of Racal Telecoms was floated on the LSE in 1988 [35], and soon the Racal Company was valued at less than its holding in Racal Telecoms.

Inevitably, Racal Telecom was demerged in 1991 and renamed Vodafone. Gerry Whent, the CEO at the time of its demerger, handed over to Christopher Gent in 1997 and the company continued to expand. With its strong track record and investor rating, Vodafone could acquire mobile telephony businesses throughout the world. The year 1999 was a particularly active year when it acquired an important stake in Verizon Wireless in the United States, which was created by the merger of Vodafone's U.S. assets with those of Bell Atlantic Corp. This became one of the most important mobile phone operators in the United States. In the same year, it also took over Mannesmann in Germany in a rare major contested takeover that was finalized in 2000 at a cost of £101 billion [36].

While the creation of a U.K.-based, world-leading, mobile phone operator is impressive, it must be questioned whether the U.K. telecoms supply industry benefited much from this. The Vodafone prospectus quoted that £62.5 million was spent on Ericsson equipment in the previous 5 years representing 40% of Vodafone's total capital expenditure. Racal did start a joint venture with Plessey to create Orbitel, which was developing base-station equipment using technology licensed from Ericsson and Matra (of France), but this did not become a significant supplier.

From its base in Newbury, Berkshire Vodafone remains a major player in the world's mobile telephone business. By 2000, it had operating businesses in many countries in Europe, Asia, and the United States with a reported 39 million users of its networks in 2000 [36]. Because it was founded by a company outside the mainstream U.K. telecoms industry suppliers, it is hard to see any great advantage that the creation of this Tier 1 mobile network operator conveyed on the U.K. supply industry.

Having spawned the huge Vodafone business, the history of the rest of Racal was a bit of an anticlimax, with its key assets being steadily divested in the 1990s. In 2000, the residual Racal business was acquired by the French Thomson-CSF group for a consideration of around £1 billion.

References

[1]   Wu, T., *The Master Switch: The Rise and Fall of Information Empires*, New York: Random House, 2010.

[2]   Brock, G. W., *The Telecommunications Industry: The Dynamics of Market Structure,* Cambridge, MA: Harvard University Press, 1981, pp. 89–109.

[3]   Wu, T., *The Master Switch: The Rise and Fall of Information Empires*, New York: Random House, 2010, pp. 45–60.

[4]   Wu, T., *The Master Switch: The Rise and Fall of Information Empires*, New York: Random House, 2010, pp. 104–107.

[5]   Capron, W. M., *Technology Changes in Regulated Industries,* Washington D.C.: Brookings Institute, 1971.

[6]   Lazonick, W., and E. March, "The Rise and Demise of Lucent Technologies," *Innovation and Competition in the Global Communications Technology Industry (INSEAD),* Fontainebleau, France, August 23–24, 2007.

[7]   Wu, T., *The Master Switch: The Rise and Fall of Information Empires*, New York: Random House, 2010, pp. 245–249.

[8]   Noam, E., *Telecommunications in Europe,* New York: Oxford University Press, 1992, p. 141.

[9]   Noam, E., *Telecommunications in Europe,* New York: Oxford University Press, 1992, p. 85.

[10] Noam, E., *Telecommunications in Europe,* New York: Oxford University Press, 1992, p. 252.

[11] L. M. Ericsson, *Annual Report,* Stockholm: Ericsson, 2001.

[12] Brock, G. W., *The Telecommunications Industry: The Dynamics of Market Structure,* Cambridge, MA: Harvard University Press, 1981, pp. 98–101.

[13] Nussey, S., and M. Yamazaki, "Japan's NTT Launches $40 Billion Buyout of Wireless Unit Docomo," Reuters, September 29, 2020.

[14] Aris, S., *Arnold Weinstock and the Making of GEC,* London, Aurum, 1998.

[15] Crowe, B., et al., *Weinstock: The Life and Times of Britain's Premier Industrialist,* London: HarperCollinsBusiness, 1998.

[16] Crowe, B., et al., *Weinstock: The Life and Times of Britain's Premier Industrialist,* London: HarperCollinsBusiness, 1998, pp. 67–69.

[17] Crowe, B., et al., *Weinstock: The Life and Times of Britain's Premier Industrialist,* London: HarperCollinsBusiness, 1998, pp. 61–67.

[18] *Statement on the Future of the Telecommunications Industry,* London: Trades Union Congress, 1977.

[19] Harris, D., "United We Stand," *Times,* December 4, 1985.

[20] U.K. Monopolies & Mergers Commission, *General Electric Company & Plessey Company,* London: HMSO Cmnd. 9867, 1986.

[21] Hemp, P., "Britain's GEC, Plessey Tentatively Plan to Join Telecommunications Operations," *Wall Street Journal*, October 17, 1987.

[22] Mathieson, C., "Marconi: The Gambler Versus the Miser," *Times*, July 6, 2001.

[23] Rees-Mogg, W., "Marconi Is the Victim of Greed and Stupidity," *Times*, July 9, 2001.

[24] Ritchie, B., *Into the Sunrise: A History of Plessey 1917–1987,* James & James, 1989.

[25] Malerba, F., *The Semiconductor Business,* Madison, WI: University of Wisconsin Press, 1985, p. 69.

[26] Simpson, D., "Telecom Delays Add to Plessey's Woes," *Guardian*, May 24, 1985.

[27] "Making a Connection at Last," *Times (London)*, October 2, 1987.

[28] Skapinker, M., "The Great Divide of Plessey Spoils," *Financial Times*, April 4, 1990.

[29] Young, P., *Power of Speech*, London: Allen & Unwin, 1983, p. 52.

[30] Bolger, A., "STC Agrees £1.9B Takeover Bid from Northern Telecoms," *Financial Times*, November 9, 1990.

[31] Bob's Old Phones, "The Telephone Manufacturing Company of Britain," www.telephonecollecting.org/Bobs phones/Pages/TMC/TMC.htm.

[32] "Victory for Pye," *The Economist*, July 23, 1960, p. 388.

[33] h2g2, "A Short History of Telecommunications in Malmesbury," 2010, https://h2g2.com/edited_entry/A60814451.

[34] Feder, B. J., "Racal Electronics: A British Success," *New York Times*, December 15, 1982.

[35] "Vodafone Prospectus," *London Times*, October 13, 1988, pp. 33–36.

[36] *Report and Accounts Year Ending 31 March 2000,* London: Vodafone Airtouch plc, 2000.

6

Conclusions

Having looked at what happened with the U.K. telecoms industry in the years since 1950, it is possible to draw some basic conclusions:

1. How did the industry perform?
2. Was this result inevitable?
3. What were the main factors causing the result?
4. What lessons can we learn from these results?
5. What are the implications for U.K. business and economic policy?

6.1 Performance of the U.K. Telecoms Supply Industry

As shown in Appendix A, in the period from 1950 to 2000, the United Kingdom went from being one of the biggest exporters of telecoms equipment to being a very large importer of telecoms equipment. This was in a period when the overall market for equipment spectacularly increased to a worldwide demand of around £100 billion per annum.

This performance contrasts poorly with some other advanced economies, as shown in Appendix A. Looking at major developed countries in Tables A.1 and A.2 (United States, United Kingdom, France, Germany, and Japan), most of these countries saw their trade balance in telecoms deteriorate in the period. This is mainly due to the increasing demand for peripheral equipment (such as handsets, routers, PABXs), which over the period became commoditized and mainly supplied by lower-cost suppliers in countries such as China and South Korea. However, the United Kingdom and the United States saw a much bigger trade deficit develop as these countries had a weakened telecoms supply industry that enabled greatly increased imports into their market and a lack of compensating export sales. While this analysis has centered on the development of electronic switching for the fixed-line telephone network, which was the key issue in the 1970s, the market had moved on by 2000, with most of the business for telecoms suppliers coming from mobile phone networks and data communications. However, while some Tier 1 suppliers arrived with innovative products after the 1970s (Nokia with mobile handsets and Cisco with data communications), most Tier 1 suppliers in 2000 were those that successfully developed fixed-line electronic switching in the 1970s.

By most standards, the U.K. performance can be seen as an economic failure. If the United Kingdom had performed closer to the average of the countries shown above, then this would have increased the balance of payments by several billions of pounds per annum in 2000 and GDP by over £50 per head and 20,000 more jobs.

6.2   Was This Result Inevitable?

There is no fundamental reason why the performance had to be so poor. The United Kingdom remained a significant economy over the period with good financial resources. The telecoms supply industry had a strong technical base on which to draw and a captive market in the United Kingdom on which to develop its products.

Tables A.1 and A.2 include Sweden, although this is not regarded as a major developed economy (its GDP being around 15% of the United Kingdom in 2000). What the data for Sweden shows is what the creation and support of a major Tier 1 supplier to the telecoms (namely, Ericsson) can do for a nation's trade balance. The United Kingdom could have seen a similar benefit if it had been successful in supporting its own local telecoms supply industry.

While some factors worked against the United Kingdom (such as its Strowger installed base), there were other factors in the United Kingdom's favor. In particular, U.K.-centered firms started the period as the main supplier to a host of territories beyond the United Kingdom (such as India, Australia, and many parts of Africa), which could have served as a springboard for rapid growth. The United Kingdom's failure was in two main aspects: the failure to successfully develop core products for its home market, and the failure to organize an effective export activity to benefit from the worldwide growth in telecoms.

6.3   What Were the Main Factors Causing the Result?

In many ways, the United Kingdom's performance was a "perfect storm" in which several apparently unrelated factors combined to produce an unfortunate result that, even with hindsight, it would have been difficult to avoid. I would identify key factors as:

1. *Technical leadership:* The way that the United Kingdom was organized with the monopoly telecoms service provider (GPO) having full design authority and regarding the telecoms supply industry as little better than "cost-plus" suppliers. This attitude pervaded even after the privatization of what became BT. Even in comparable situations in countries such as France and Germany, a healthier relationship seemed to operate.

   The attitude of GPO set the culture for the entire industry and survived into the twenty-first century. The management of the key U.K. suppliers GEC, Plessey, and STC was extremely reluctant to engage in serious investment in new products outside the scope of GPO-sponsored developments (such as System X). Hence, when a major new market opportunity presented itself to these companies (such as mobile telephony), their engagement was totally inadequate to match the scale of the market opportunity, even though they processed the technical capabilities, the manufacturing capabilities, and the management resources to succeed.

2. *Export organization:* Nowhere have I seen any sign that GPO regarded itself as seriously involved in the export of U.K. telecoms equipment. As telecom networks became more complex, system design became a vital part of the sale of telecoms equipment. The U.K. suppliers were at a considerable disadvantage compared with countries such as Sweden and France, which ensured that its exporters could offer this capability. The creation of an industry-owned sales organization (BTS) in 1979 was too late and never properly supported by its owners. It was only when GEC eventually gained full control of GPT in 1989 that some effective export marketing efforts were made to export System X. This was a classic example of too little, too late.

3. *Upgrade strategy:* The period started with the widespread and continuing use of obsolescent technology (Strowger-based switching) within the United Kingdom. The approach to upgrading the network proved to be poorly conceived. This was a result of GPO's mindset that can be traced back to the 1920s. While the continuation of Strowger in production right up to the 1970s gave the U.K. government a convenient

way to support employment in economically depressed areas, it also delayed the introduction of other switching solutions that may have prevented some of the factory closures during the 1980s and 1990s. The U.K. supply industry was not able to use the United Kingdom as an effective test-bed to validate new products that were in growing demand in export markets. In particular, because of the delay in adopting interim switching solutions (using Crossbar or reed relays), there was little experience to offer export customers a migration path towards the eventual full digital switching solutions. Similarly, the culture of the Thatcher era of the 1980s meant that, in the United Kingdom, the rapid rollout of the mobile phone network using private money dominated the plans with little thought as to how the U.K. telecoms supply industry could benefit from this enormous market opportunity.

4. *Technological arrogance:* There is a certain arrogance in the U.K. attitude to technological change. The United Kingdom has a strong track record of innovative thinking but a relatively poor track record of implementing major new technologies. Thus, GPO delayed investing in interim solutions to its switching needs because it expected full digital solutions to become available soon. However, it did not seriously think through the evolution path needed to upgrade the network. Not only did this affect the U.K. network, but it weakened the ability of the U.K. industry to support its export customers who were similarly looking to upgrade their networks. Similar arrogance can be seen in the System X program. Despite BT being in the privileged position of knowing the planned timescales of the Swedish Ericsson AXE exchange, the country chose to ignore these inconvenient facts and plan to develop System X against a more ambitious timescale. System X duly arrived about 5 years behind schedule and about five times the original budget.

5. *Privatization of BT:* This had a major effect on the whole dynamic of the industry. Although still controversial, this change did make a serious restructuring of the whole industry possible. However, the privatization in 1984 was at an extremely unfortunate time with regard to two major events in telecoms development: the development of a full digital switching system (System X), and the early introduction of mobile telephony. In both developments, the main driver was to satisfy end-user requirements. While laudable, this meant that the supply industry did not have enough time to adapt to the new regime in terms of both the commercial environment that it brought in and the loss of a captive U.K. market in which to develop its capabilities.

6. *Poor management:* The management of the key telecoms suppliers had a major effect. The three key suppliers operating in the 1970s and 1980s (GEC, Plessey, and STC) were all conglomerates with a substantial part of their revenue coming from outside the telecoms market. They were also run by strong, possibly brutal, leaders usually focused on short-term results. They subsequently found cooperation difficult and were not up to the task of plotting a route for the evolution of the industry. All three companies were used to product development using a cost-plus regime prevalent in defense and also in the GPO-led telecoms industry made them extremely reluctant to invest in speculative developments. A merger of the companies' telecoms interests in the 1970s would have probably given the industry a better chance of success, but this was off-limits even for an interventionist Labour government because of the personalities involved. Despite the acknowledged issue of a fragmented supply base, the United Kingdom ploughed on with the System X development to an almost inevitable conclusion.

7. *Unsuccessful mergers:* Having failed to merge operations in the 1970s, the eventual mergers in the 1990s were too late and poorly conceived. STC's merger with ICL was never likely to succeed and went into reverse within a decade. The GEC-Plessey merger was delayed and damaged by governmental concerns over the defense interests of the companies. This brought in Germany's Siemens as a partner. They, in turn, were a rather half-hearted partner (with parallel product lines). At the end of the 1990s, GEC "dash for growth" in the post-Weinstock period was another disastrous series of ill-conceived deals that left the U.K. government with a monopoly defense supplier that it wanted to avoid (British Aerospace) and saw GEC (or Marconi, as it then styled itself) make disastrous acquisitions in the United States and implode spectacularly by 2004, taking with it any prospect of a U.K. Tier 1 telecoms supplier.

Together, these seven factors listed contribute to the United Kingdom's overall performance. Several of the other countries that I have reviewed suffered from similar factors. For instance, Japan started with a weak telephone infrastructure and even embarked on a development of an electronic exchange equivalent to the United Kingdom's System X while involving a much more complex array of companies than in the United Kingdom. However, the overall culture of industry cooperation in Japan seems to have avoided many of the issues seen in the United Kingdom both in regard of the technical collaboration and exporting the developed product.

6.4   What Lessons Can We Learn?

Perhaps the main lesson is that, in line with classic economic theory from Schumpeter, even in a supposedly rapidly changing environment such as the telecoms industry of this period, long time cycles are involved with plenty of creative destruction. There is a clear mismatch between the political time cycle of 5 years and the economic cycle of approaching 50 years (an example is that although Strowger switching was accepted as obsolescent in 1950, the last Strowger exchange was switched off in 1990).

Britain's two-party political system is also a factor. A narrowly won election could have dramatic effects in industrial strategy, with the country swinging between central planning and a free market paradigm. The benefits or otherwise of these two systems can be debated elsewhere, but rapidly switching between the two approaches prevents any long-term plans being successfully adopted.

The management of the industry, particularly in the 1960s to 1980s, is certainly a factor. The key players seemed to be run as personal fiefdoms of domineering bosses. Undoubtedly, there were able and intelligent people in all the companies, but their skills were not effectively harnessed to the task of developing the industry.

While it might be tempting to denigrate the quality of management mentioned above, they operated very effectively within the financial regimes of the time. This includes the fabled "short-termism" of the City of London, which was the dominant source of finance for these enterprises. Any managers announcing that they fancied investing serious money in new speculative investments (such as mobile phones) could easily see the value of their company tumble and the consequent takeover by an asset stripping corporation.

The loss of technical leadership in telecoms could be simply regarded as an unfortunate but minor issue were it not for the fact that similar factors can be seen at work in many other industries, including automobiles, aerospace, semiconductors, computers, and motorcycles. Perhaps even more worrying is that current "successes," such as pharmaceuticals, could become part of this list.

Does this matter? If this book had been written immediately after 2000, then we could have looked at industry champions such as Nokia, Ericsson, Nortel, and Alcatel as examples of good practice. However, within 15 years, many of these companies had lost their preeminent position mainly to Asian competitors such as Huawei. So perhaps nothing is forever and success is transient. However even if good practice only adds perhaps 20 years to an industry's life cycle, this can be very beneficial to a country's economy if this repeats over many industries.

I believe that there are two key issues addressed by the book. The first is the United Kingdom's need to identify key strategic industries and support them. The second is the need to achieve far better performance from new, technology-led business opportunities.

On the first issue, this book starts with the rather ludicrous situation of BT being forced to rip out recently acquired equipment from Huawei to replace it with equipment from "friendlier" (but still foreign) equipment from the likes of Ericsson. If anyone thinks that poor appreciation of the strategic importance of certain industries is a recent phenomenon, then please read a description of what happened when the United Kingdom failed to support its optical industry over 100 years ago [1]. The article could have equally cited the United Kingdom telecoms industry as another example. By failing to nurture key strategic industries over a long period, the country will always be faced with making difficult and expensive procurement decisions when its national interests are at stake. The recent Brexit decision exacerbates this issue as, if Brexit is to mean anything, I guess it is about independence. This is a rather empty concept if the United Kingdom is going to be forever beholden to other countries on strategic matters such as the availability of "Gigaplants" making batteries for electric vehicles. Recent experience with the Covid-19 pandemic and the switch to green energy indicated that observers will never be short of examples of this long-running phenomenon. I believe this is a feature of the way in the United Kingdom that the political establishment interfaces with its executive arm (the civil service).

The second issue relates to economic performance. This book lays out the long-term economic performance of the United Kingdom since the industrial revolution, which was based on world-leading industries of textiles, coal, steel, and shipbuilding. Throughout the twentieth century, the United Kingdom struggled to come to terms with the decline of these industries (while fighting two costly major wars), but this was largely cushioned by the large reserves of wealth held in assets such as overseas companies. This wealth has now largely been exhausted and the United Kingdom has shown a deficit in both its trade balance and its fiscal balance for most of the last 50 years. Meanwhile, since World War II, the United Kingdom has developed an increasing urge to live beyond its means. The postwar push for a "land fit for heroes" spawned the National Health Service and other very worthy social programs. The growing expectations of the population for better health care and living conditions drives social spending costs upwards. Meanwhile, the legacy of the Thatcher years has put an effective cap on the taxation burden that is politically expedient. To date, this gap has been bridged by increasing borrowings and the sale of national assets. However, the United Kingdom will run out of things to sell and perhaps its creditors will decide it is overborrowed. The U.K. government's own Office for Budget Responsibility (OBR) produced a report in 2023 that extrapolated the United Kingdom's economic performance for the coming

50 years and produced projections of national debt into completely uncharted territory at 3 times the GDP [2].

To overcome this issue, the United Kingdom aspires to be primarily a knowledge-based economy. If this is the case, then it needs to get much better at exploiting this knowledge for the benefit of its citizens. In this book, I have explored in some detail how the opportunities offered by one particular knowledge-based industry have largely passed the United Kingdom by. As mentioned earlier, similar analysis could be applied to a whole list of similar situations. The future awaits us with a host of new opportunities such as green energy, artificial intelligence, and life sciences. In all these areas, the United Kingdom has the position to successfully exploit the opportunity. Failure to do this will leave the United Kingdom in a downward spiral where its only way to meet the growing financial aspirations of its population is by continuing to "sell off the family silver" and becoming subservient to other countries and organizations, which we find unattractive. At some point, the mismatch of the population's aspirations and the establishments' ability to deliver will cause social tensions and the sort of instability that we have witnessed elsewhere.

Perhaps the most pertinent recent example of a current market opportunity similar to that posed by telecoms in the 1960s is green energy. Here there is a general acceptance of the enormous opportunity, and this has been reflected in numerous government pronouncements on the subject. However, the rhetoric of these announcements compares poorly with the actions that the country is undertaking to exploit this opportunity. So in the next 10 years, the whole program is in danger of being stalled by the failure to roll out sufficient infrastructure and skilled workforce to support the rhetoric.

This analysis could make a call for dramatic changes in the U.K. political/cultural establishment. However, this is a country seeped in history that does not easily embrace change. It cannot even agree on any serious reform of an unelected second chamber of government, with almost 1,000 members comprising hereditary peers, bishops, and beneficiaries of political patronage. The United Kingdom's multiethnic population is still expected by the establishment to feel honored to be awarded the title "Member of the British Empire." The United Kingdom is unlikely to embrace radical change without some cataclysmic event (such as losing a major war).

It is not all negative. The United Kingdom has achieved a strong position in some new industries, despite them not receiving large direct support. Examples are computer gaming and the media sector. These are major contributors to the U.K. economy. In many ways, these sectors suit the U.K. culture of craft-based cottage industries. Support for the media sector is, in effect, extensive but wrapped up in the U.K. establishment with money coming through publicly funded broadcasting (BBC) and arts funding (Arts Council). Computer gaming (a more than billion-pound industry) seems to have simply failed to be noticed by the political establishment.

The United Kingdom faces a difficult future if it is forced to rely on the revenue of those industries that fit into its culture coupled with a continuing program of "selling the family silver" (over 50% of houses in London worth over £1 million are being sold to nonresident expatriates) [3].

6.5 Implications for U.K. Business and Economic Policy

While it is tempting to advocate for the complete restructuring of the U.K. political and business system, this is not my intention. Many writers have provided excellent checklists of the sort of changes needed for British society and industry to perform better. Some of these have been cited earlier [1]; another example covers the reality of how poorly managed and capitalized the U.K. engineering industry was at the time [5].

Radical change of that nature has unpredictable consequences, and there can be no guarantee that such changes will have an aggregate positive outcome. The U.K. historic model of slow but steady minor changes has provided a stable environment for the improving prosperity and well-being of the population (give or take a few world wars).

My general suggestion would be: Agree what we can agree on, provide resources and infrastructure for development, and let the market do the rest. If we take those three factors in order and relate them back to the telecoms industry history:

1. "Agree what we can agree on": Even in 1950, there was a considerable consensus on several factors in the industry, notably:

   • The GPO was bureaucratic and needed reform.

   • There would be a need for substantial upgrade and investment in the network.

   • There were too many suppliers who were operating inefficiently.

   This could have been an agreed government policy that could gain cross-party support. In practice, it took too long to even start to address these issues. Basically, the issue was not high on the political agenda.

   Today there seems to be a clear need for a political consensus on investment issues such as green energy, as well as major expense items such as health care.

2. "Provide resources and infrastructure for development": Eventually these issues were addressed, but in a rather belated fashion:

- On a national scale, substantial growth in university education (starting in the 1960s) has increased the technical skill base of the industry, while a later growth in business schools (starting in the 1970s) led to some improvement in management. There was little appreciation at the key time (1960s to the 1980s) of the strategic importance of telecoms. Nor was there any understanding of the difference in nature of telephone service providers (BT) and telephone equipment suppliers (GEC, Plessey). The Labour government of the 1960s talked the talk about the "white heat of technological innovation," but could have actually started the restructuring of the industry 20 years earlier than it happened.

- The separation and eventual privatization of BT in 1984 made it possible to raise the necessary funds for the network development. Similarly, the use of a full private capital approach to the creation of the mobile network led to its successful deployment.

- The success of Racal in creating and then spinning off Vodafone shows that, despite the legendary short-termism of the London Stock Exchange, a good story well told could raise significant funding. The telecoms supply industry needed such a champion.

What remains of the U.K. telecoms industry is largely due to the accepted excellence of the technical resources in the country. This might be a diminishing factor without large players in place to train the next generation. However, such skill sets do endure for decades.

I hope that the improvements in resources will lead to better industry performance when faced with market opportunities. However, the current failure of the government to train enough skilled craftsmen (builders, plumbers, electricians) to roll out green energy does not bode well.

3. "Let the market do the rest": By this, I do not mean the unchecked power of the free market be allowed. To a considerable extent, the fate of the telecoms supply industry after 1980 was left to the free market and the results were recorded in this book.

The world in which we live does not have many level playing fields, and I feel that a light-touch free market is appropriate:

- The U.K. market should be accessible to U.K. businesses to develop their offering before they are put on the world stage.

- The value of access to the U.K. market should not be underestimated and a good price should be extracted from overseas suppliers for this access in terms of jobs created and location of key design and management facilities.

- Similarly, the United Kingdom prides itself on its aid budget. Perhaps more of this can be offered in goods and services to help countries build infrastructure.

- Increasingly, the City of London is being bypassed as a funding route for enterprises. This, in turn, may reduce the short-termism attributed to some finance.

This is not a new message; it is remarkably aligned with the conclusions of Freeman in his work on innovation, first published in the 1970s [6]. However, deriving conclusions that are consistent with those postulated almost 50 years ago does not make them less relevant. Whether this approach will be adequate to meet the population's expectations for a continuing improvement in their collective wealth and well-being is debatable, but it seems to be all that is available.

However, my motivation in writing this book is primarily to chronicle a chapter of history of the United Kingdom, which will be easily lost. The factories that once supplied the U.K. telecoms industry have mainly disappeared and have been repurposed as superstores and the like. Perhaps if people knew more about this history, they would not repeat the same mistakes? However, the only lesson of history is that humans never learn from it.

References

[1]   Conway, E., "Our Race to Embrace Electric Cars Has Let China Get Ahead," *Sunday Times*, July 8, 2023.

[2]   Martin, B., "Debt, Sickness and Power Sap UK's Finances, OBR Finds," *London Times*, July 14, 2023.

[3]   Hill, D., "London Housing Crisis: How Far Are Super-Rich Foreigners to Blame?" *Guardian*, October 22, 2013.

[4]   Hutton, W., *The State We're In*, London: Vintage, 1996.

[5]   Brown, T., *Tragedy & Challenge*. London: Matador, 2017.

[6]   Freeman, S., *Economics of Industrial Innovation*, Oxford: Routledge, 1997.

Appendix A
Telecoms Industry Import/Export Statistics

It initially seemed a relatively simple task to introduce tables showing the performance of the telecom supply industry of key countries. However, reality intervenes in several ways:

1. As technology and the industry evolve, there is a continuing need to decide what fits inside or outside the definition of items. For instance, "should a mobile phone handset (which was not even conceived of in 1950) be included in the statistics for telephone systems"? Similar discussions can be had about telecoms attachments including modems, routers, and fax machines. For the purpose of the analysis in Tables A.1 and A.2, I have concentrated on the telecoms network and tried to eliminate the effect of supply of telecoms attachments as described above.

2. As described in the introduction, the period of study spans the transition of paper-based data to computer-based data. This watershed happened around 1970 and, in practice, most accessible data sources have little information available before this data. In practice, this is not a major setback because the main trends in the telecoms industry were driven by the application of digital switching that took place mainly after this date.

Table A.1
Trade Statistics for SITC2_7641

U.S. Millions of Dollars		1978	1980	1985	1990	1995	2000
France	Imports	38	61	57	205	675	1,879
	Exports	124	187	234	561	1,504	2,493
Germany	Imports	69	94	94	808	1,971	3,221
	Exports	528	642	470	1,089	2,390	2,216
Japan	Imports	18	21	57	420	1,190	2,595
	Exports	359	397	1,379	3,306	2,948	1,999
Sweden	Imports	20	35	41	197	309	535
	Exports	381	478	515	873	435	3,962
United Kingdom	Imports	64	82	228	599	2,010	4,487
	Exports	118	126	181	519	1,970	6,103
United States	Imports	239	432	2,099	2,610	3,695	12,715
	Exports	388	557	832	1,358	3,248	6,314

Table A.2
Adjusted Trade Statistics

U.S. Millions of Dollars		1965	1970	1980	1990	2000	2005
France	Imports	24	45	187	311	1,172	1,451
	Exports	38	57	375	812	2,205	1,840
Germany	Imports	34	84	277	373	1,523	3,020
	Exports	105	210	810	1,404	2,819	3,504
Japan	Imports	4	18	53	460	3,335	1,724
	Exports	137	467	856	1,558	4,268	1,379
Sweden	Imports	19	44	64	181	816	808
	Exports	28	75	544	1,371	4,619	3,452
United Kingdom	Imports	25	59	240	694	3,856	5,192
	Exports	77	111	347	612	1,433	2,583
United States	Imports	105	367	463	1,246	8,429	16,996
	Exports	118	225	1,033	1,964	8,085	9,517

3. The data reported is primarily import and export and rarely covers domestic production and consumption. However, the pattern of import and export gives a reasonable indication of the health of the domestic supply industry.

4. Over the period of this study, industry classifications have themselves moved. Several UN organizations collect data (such as the OECD), with most nations reporting information to this body from the 1960s, but over that period to now, the classification codes have changed many times. Between 1960 and 2010, no less than nine different classification systems were in use. These evolved to provide ever-increasing granularity in reporting as the classification systems struggled to keep up with economic trends. The classification codes fall into two main systems: the Standard International Trade Classification (SITC) and the Harmonized System (HS), which evolved in parallel with revisions every 8 years or so.

5. Given the more than 150 nations reporting trade data, it is inevitable that there will be inconsistencies and inaccuracies in reporting between countries. This is particularly true in products such as mobile phone handsets where the data classifications did not clearly state what code to use until well after significant trade had been established.

6. However, even these statistics are not reliable. An example of these is how the U.K.'s telecoms trade balance was massively distorted by "carousel fraud" in the period around 2000. In this, items were wrongly declared in one jurisdiction before being passed on to another country with the benefit of value-added tax (VAT) refunds going to criminals [3]. In the one reported case that I have cited, a single fraud produced around a £2 billion wrongful declaration of trade in 1 year, which undoubtably found its way into the U.K. trade statistics. Mobile phones were often the chosen commodity for this sort of fraud [1], having a high value-to-weight ratio and being massively traded. Thus, export trade reported for telecoms products from the United Kingdom in 2000 greatly exceeds the known production of U.K. telecoms suppliers and is dominated by the import and export of goods, often bogusly declared. This phenomenon is unlikely to have been unique to the United Kingdom as the single market in the European Union (at that time) did not have unified VAT rates, but did have an unrestricted flow of goods. Clearly, we must remove these aberrations to get meaningful data.

7. The increasing globalization of the telecoms market has already been discussed. One result of this is the increasingly complex supply chain to satisfy telecoms systems orders. This is often reflected in increased import/export numbers as products are moved in and out of countries throughout the world to satisfy a particular order.

8. Even when data is correctly reported to UN bodies, their accurate assembly into statistics are not guaranteed. Thus, two reporting UN

bodies can sometimes report the same trade information, using the same classification system, and produce slightly different numbers.

9. Over the period of my interest, I have produced a data set that is my best estimate of core telecoms products (excluding handsets and telecoms attachments as described in item 1). This is based on the SITC code series. For instance, for SITC_7641: the summary definition (for Rev3 of the Classification) operative from 1990 is "Telephone equipment which includes Telephone sets, Teleprinters, Telephone switch equipment, Telephone line equipment and other Telephone equipment." From this, I have adjusted for handsets and other attachments. However, for earlier data, I have had to use earlier revisions of the SITC classification, which were less precise and hence required further adjustment.

Tables A.1 and A.2 are my personal workings of the data which I use for discussion within this work. First, Table A.1 gives the raw OECD statistics for SITC2_7641 for the six key countries in U.S. dollars. I review France, Germany, Japan, Sweden, the United Kingdom, and the United States.

To extend the data back to earlier times, one must use several sources that are not always consistent between countries and include an element of estimation to fill in gaps in the data. Doing this and eliminating known aberrations in the statistics (such as the carousel fraud mentioned earlier) gives the data in Table A.2.

Having generated a table that shows the trade performance of the key telecoms players over the period, it is now possible to review this in line with the commentary already made in Chapter 5. To show how these trends continued, I have included data for 2005. The crucial information in the table is the net trade balance of each country, which is less likely to be affected by the reporting anomalies previously described.

A.1 France

As described, the French telecoms industry started in a relatively weak state, and this is reflected in the early trade statistics. Like the United Kingdom, France benefited from export sales to its former colonies. The state-driven investment in telecoms infrastructure from the 1970s stimulated export growth. Thus, Table A.2 shows a $1 billion trade surplus in 2000. However, while the acquisition of the ITT telecoms businesses in Europe and the later acquisition of Lucent in the United States made Alcatel a leading telecoms supplier in the 2000s, much of this revenue was generated from territories outside France. The results are consistent with those reported by Thatcher comparing U.K. and French exports and imports from 1970 to 1995 [2].

A.2 Germany

Table A.2 shows that Germany achieved a reasonably favorable export performance in early years. This would seem to be mainly achieved by the strong export capabilities of Siemens and SEL. While a significant part of the telecoms industry was in foreign hands from the 1990s (notably the French acquisition of SEL), a reasonable balance of trade was maintained with a trade surplus of around $700 million in 2000.

A.3 Japan

Rather like Germany, the inherent strength of the main Japanese telecoms manufacturers (such as NEC) in exporting supported a continuing balance of payment surplus despite Japan largely falling behind in technology (particularly in mobile phones). Thus, a nearly $1 billion trade surplus was recorded in 2000. However, by 2005, strong competition from other Asian manufacturers, particularly in the supply of telecoms attachments such as mobile phones, had created a trade deficit.

A.4 Sweden

Sweden is a much smaller economy than the others listed. However, its performance is noteworthy. It is the only country shown to maintain a strong export surplus through the period shown (1965–2005). This is almost entirely due to the performance of one company, L. M. Ericsson. Ericsson managed to maintain a strong technical leadership in the telecoms industry, which dated back to the 1930s. While it faced several setbacks (such as being excluded from the U.K. market from 1948 to 1968 under the deal to sell off its U.K. Beeston factory), it got many of the big decisions right. In particular, its involvement in the early Nordic trials for mobile telephony gave it a strong position in mobile telephony networks. While it had to eventually withdraw from making mobile phone handsets, Ericsson is still one of the top telecoms equipment suppliers, with over half its revenue in 2000 coming from mobile telephony equipment [3].

A.5 The United Kingdom

The United Kingdom's performance is discussed in more detail in the remaining appendicies. What the data shows is that the country managed to delude itself about its telecoms situation for most of this period. Despite claims made by government advisors in the 1970s that the United Kingdom was the world's biggest telecoms exporter, this was probably not true from 1950. Due to its failure both to develop competitive products and achieve a coherent long-term export marketing strategy, the country was in trade deficit by the 1980s from which it has never emerged. Thus, a $2 billion trade deficit has been shown for 2000.

A.6   The United States

The U.S. performance is similar to that of the United Kingdom. The huge domestic market for equipment in the United States meant that not much attention was paid to exporting after AT&T divested most of its international activities to form ITT in 1925. The breakup of AT&T and the poorly managed early implementation of mobile telephony made the United States vulnerable to import penetration, but the biggest effect of this occurred after 2000 with the fire sale of Lucent to Alcatel in 2006. However, unlike the United Kingdom, the United States could weather the trade deficit and the development of the internet and social media were game changers that enabled the U.S. bounceback. However, the possible dependence of the United States on foreign suppliers of telecoms infrastructure has started to become a major issue.

References

[1]   The Guardian, "Carousel Fraud Ringleader Jailed," *The Guardian,* July 8, 2012.

[2]   Thatcher, M., *The Politics of Telecommunications: National Institutions, Convergences, and Change in Britain and France,* Table 9, Oxford: Oxford University Press, 2000.

[3]   Ericsson, Accounts for 2000, Annual Reports, Stockholm, Sweden, 2001.

Appendix B
Telecoms Factories in the United Kingdom

The U.K. telecoms industry locations reflect the socioeconomic conditions of the country in the twentieth century. Most of the locations are the result of decisions made well before 1950 and the sites are a clear representation of phases that can be seen in other engineering industries, notably the car manufacturing industry.

- *Early manufacturing in key industry centers:* In the case of telecoms, this was London (Woolwich), Nottingham (Beeston), and Liverpool.
- *Move to better locations:* This is particularly true of businesses started in London that migrated to sites on the outskirts of London such as Southgate and Basildon as their scale grew. After starting in Liverpool, the GEC company chose Coventry as its main site.
- *Satellite factories in depressed areas:* The U.K. government put a lot of effort into finding work to replace declining industries (coal, steel, shipbuilding) offering large financial incentives for firms to relocate. Normally, telecoms suppliers responded by setting up satellite factories in the areas such as Northwest England (Merseyside), Northeast England (Tyneside, Teeside), South Wales, Central Scotland, and Northern Ireland. These provided relatively low-grade employment but lots of it (in the 1960s and 1970s, when Strowger was in full production). The importance of this program to the regions is illustrated in Figure B.1 by the

Figure B.1   *Financial Times,* May 27, 1983. (Source: Invest Northern Ireland.)

advertisement opposite, placed by the Northern Ireland Development Board.

However, the switch to electronic exchanges led to a closure of most of these facilities starting in the mid-1980s, just as the Thatcher era was shutting down the traditional industries such as coal and steel that they were supposed to replace.

• *Move to pleasant suburbs:* As the industry became knowledge-intensive, then there was a need to build up high-grade technical teams. These graduate electronic/software engineers demanded more agreeable locations, so many sites were repurposed for their development. This would include places such as Plessey Poole (Dorset) and Racal Newbury (Berks) whose only real advantage was being a nice place to live reasonably close to London.

Table B.1 lists all the major sites involved in telecoms work in the period 1950 onwards.

The 35 sites listed in Table B.1 made up most of the U.K. telecoms industry activity in the period of 1950 to 2000. I have shown the estimated employment of each site in 1975, which was close to the high point of telecoms employment in the period towards the end of the maximum Strowger output era and before the lower manpower needs of electronic exchanges came into play. The numbers are largely gathered from the TUC submission to government [2], which included their survey of employment in an attachment. However, it should be realized that in addition to the number of around 56,000 employees that they listed for these sites, you must add around 20,000 more employees that they listed from the main suppliers who at the time were operating outside these sites (mainly supporting telephone exchanges). The TUC survey is also silent about staff who were predominantly engaged in developments leading into System X (which started in 1977). As mentioned earlier, these were largely distributed over around 10 or more telecom sites, including:

- *BT:* Martlesham;

- *Plessey:* Liverpool, Beeston, Taplow, Poole, Roke Manor;

- *GEC:* Coventry, Chelmsford;

- *STC:* New Southgate, Basildon.

However, these employment numbers should also be compared with the telecoms employees of the GPO, as it then was. The numbers employed by BT/GPO declined from 232,000 in 1970 to 127,000 in 1996 [2].

This list omits several manufacturing activities relating to telecoms. This includes cable supply and telephony handsets. Cable supply came from several factories within the GEC and STC organizations plus some other suppliers (such as BICC). There were several specialist factories making telephone handsets in the period. BT had its own manufacturing activity in Cwmbran in Wales making and refurbishing telephones.

As the telecoms market evolved, several sites were established relating to importing technology from other countries. In some cases, these were relating to the supply of simple peripheral hardware that attached to the network such as handsets, fax machines, and small PBXs. Thus, suppliers from Japan and elsewhere established locations from the 1970s. In other cases, much more significant operations were established. Examples of this are Motorola and Ericsson.

Motorola, from the United States, established a site in Basingstoke, Hampshire, which was focused on the supply of radio equipment for the mobile phone industry. This site operated from 1986 to 2018 and supplied products well beyond the United Kingdom.

Table B.1
U.K. Telecoms Sites

	Region	Opened	Closed	Products	Staff 1975	Notes
GPO/BT						
Martlesham	South	1975	Open	R&D	2000	Moved from Dollis Hill, London
GEC						
Aycliffe	Northeast England	Before 1950	1990	Str	1650	Press coverage of closure, 900 jobs in 1989
Chelmsford	South	1898	Open	Trans	200	Multiproduct site, now part of BAE
Coventry	Midlands	1920	2008	Cb/Str/TXE/trans	11,500	GEC telecoms HQ site
Glenrothes	Scotland	Before 1950	1990s	Cb/Str	1,500	Siemens/AEI factory
Hartlepool	Northeast England	Before 1950	1980s	Str	3,250	Siemens/AEI factory
Kirkcaldy	Scotland	Before 1950	1980s	Cb/Str	1,500	Siemens/AEI factory, later semiconductors
Middlesbrough	Northeast England	1949?	1984	Str	1,000	0
Treforest	Wales	Before 1950	1988	TXE	500	Siemens/AEI factory
West Chirton	Northeast England	Before 1950	1980s	Str	300	Siemens/AEI factory
Woolwich	South	1863	1968	Str	0	Main Siemens/AEI factory, closed by GEC
Plessey						
Ballynahinch	Northern Ireland	Before 1950	0	Cb/TXE	300	0
Beeston	Midlands	1901	2008	Str/TXE/trans	7,200	Originally National Telephones, then Ericsson HQ
Chorley	Northwest England	Before 1950	0	Str	0	Leader in supply of payphones
Huyton	Northwest England	Before 1950	0	Str/TXE	300	0
Kirkby	Northwest England	Before 1950	1977	Cb	520	0
Liverpool	Northwest England	1903	2006	Cb/Str	9,180	Originally ATM HQ, main Plessey exchange site
Poole	South	0	0	R&D	0	Mainly Plessey controls site
Roke Manor	South	0	0	R&D	0	Mainly Plessey defense site

Table B.1 (continued)

	Region	Opened	Closed	Products	Staff 1975	Notes
Plessey						
South Shields	Northeast England	1960	1984	Cb	1,700	EU grant 1978
Sunderland	Northeast England	Before 1950	1977	Cb	2,750	0
Taplow	South	0	0	R&D	0	Telecoms development site
Wigan	Northwest England	Before 1950	1984	Str	200	0
STC						
Basildon	South	1965	2009	Trans	2,000	Closed by Nortel failure
Benfleet	South	Before 1950	0	TXE	500	0
East Kilbride	Scotland	0	1977?	Str	1,000?	0
Enniskillen	Northern Ireland	Before 1950	0	Str	500	0
Larne	Northern Ireland	Before 1950	1976	Str	760	0
Monkstown	Northern Ireland	1962	2010	Cb/Str/TXE	2,600	Closed by Nortel successor
New Southgate	South	1922	2002	Str/TXE	3,500	HQ STC telecoms
Treforest	Wales	Before 1950	0	TXE	400	0
TMC/Pye						
Airdrie	Scotland	1960?	1996	PBX	1,000?	MBO by Telecom Sciences
Dulwich	South	1922	1978	Str	500	0
Malmesbury	South	1950	1986	Str	500	0
Racal/Vodafone						
Newbury	South	1985	Open	Mobile	0	Vodafone HQ

Str = Strowger exchange production, Cb = Crossbar exchange production, TXE, TXE2, and TXE4 = exchange production, PBX = private branch exchange production, trans = telecom transmission equipment, mobile = mobile telephony development, and R&D = research and development.

Ericsson, from Sweden, had a strong presence in the United Kingdom up to 1939, including operating the Beeston factory (see Table B.1) supplying telephone handsets and other equipment. However, as a result of the World War II, the Beeston factory was bought out as a separate company called U.K. Ericsson in 1948 and as part of the deal Ericsson were excluded from the U.K. market for 20 years. U.K. Ericsson was taken over by Plessey in 1961. As the newly privatized BT became disillusioned with the progress on System X development, they turned to Ericsson as an alternative supplier. This culminated in Ericsson setting up a joint venture with the U.K. electrical group, Thorn, to supply what was termed System Y (a rebranded version of Ericsson's AXE exchange). The U.K. operation Thorn Ericsson Ltd. had its headquarters in Horsham, Sussex, and its main manufacturing plant in Scunthorpe, Lincolnshire, where it fulfilled a £100 million contract for System Y placed by BT in 1985 [3]. Ericsson bought out Thorn a few years later.

The 35 sites listed in Table B.1 represent a considerable concentration of the sites involved in telecoms supply compared with pre-1950 when there were over 100 registered suppliers to the GPO. Many of them were supplying telephones and simple telephony apparatus. The sites are grouped together based on ownership in the 1970s, but many of them owe their existence to operations well before the 1950s. The 5 core sites of 1950 correlate to the then-5 key companies operating under the BSA for exchange equipment were (in date order of founding):

- *Woolwich (1863):* Founded by the Siemens family in 1863, the oldest site that even in 1950 still traded as Siemens Brothers but was a U.K. company with primarily U.K. shareholders. It was not fully taken over by AEI until 1955.

- *Beeston (1901):* Originally founded by the then-National Telephone Company to make telephone handsets, it became jointly owned with Ericsson around 1920 and then, with the buyout from the Swedish parent company in 1948, the headquarters of the independent U.K. Ericsson company until the Plessey takeover in 1961.

- *Liverpool (1903):* The Edge Lane site in Liverpool was founded by ATM to make Strowger exchange equipment. It was ATM's headquarters until the Plessey takeover in 1961.

- *Coventry (1920):* The Coventry site operated as GECs main telecoms location and the company's biggest telecom site from its foundation through to its closure in 2008 after the demise of what was by then the Marconi Company.

- *New Southgate (1922):* The main telecoms site of STC was set up on the relocation of their telecoms business from Woolwich.

All these sites had virtually ceased to operate as telecoms sites by the early 2000s. New Southgate is now a housing site. Liverpool, Beeston, and Coventry are now mixed use with some of their area as an industrial estate and even a few buildings, surviving from the telecoms past. Ironically, the site that was best preserved at the time of this writing was the Siemens site in Woolwich, which still has some of the original buildings together with areas being repurposed as a heritage location.

The push for Strowger-based growth in exchange lines in the 1960s and 1970s led to satellite operations being set up in the development areas. Often, these locations that were used were already in operation for low-scale telecoms manufacturing (for items such as handsets). Each of the BSA 1950 suppliers favored certain locations:

- *AEI:* Mainly set up plants in the Northeast of England.

- *Ericsson:* Similarly set up plants in Northeast England, which was convenient for its Nottingham base.

- *ATM:* Developed sites around its Liverpool headquarters in Northwest England.

- *GEC:* Also favored plants in Northeast England, which was convenient for its Coventry base.

- *STC:* Favored setting up plants in Northern Ireland. This was an important source of investment for Northern Ireland, as illustrated in Figure B.1.

However, this overview is modified in some cases by other locations being chosen often as a result of specific deals made available at the time concerning commercial opportunities or investment. The detailed history of each site is beyond the scope of this work but remains a fascinating topic for research for local historians and enthusiasts.

The ending of Strowger orders decimated the satellite sites. A few were repurposed into electronic manufacturing, but most were closed after the inevitable local protests, including a few work-ins. Thus, of the 35 sites in Table B.1, 20 were in development areas of northern England, Wales, Scotland, and Northern Ireland. Of these, only a couple survived beyond 1990.

The southern suburban sites fared better, but most were later closed as a result of the takeover of Plessey by GEC and the subsequent failure of GEC (Marconi) together with the failure of Nortel, which took over STC. Thus, by 2010, the only telecoms sites existing in the United Kingdom were the residual site at Martlesham (whose vast site has been repurposed as an industrial estate including BT) and a few locations occupied by foreign-owned suppliers, notably Ericsson, which took over most of the GEC (Marconi) activities, and Siemens, which gained some Plessey assets.

References

[1]   TUC, Statement on the Future of the Telecommunications Industry, London: Trades Union Congress, 1977.

[2]   Thatcher, M., *The Politics of Telecommunications: National Institutions, Convergences, and Change in Britain and France*, Table 9, Oxford: Oxford University Press. 2000.

[3]   The Guardian, "Carousel Fraud Ringleader Jailed," *The Guardian*, July 8, 2012.

[4]   Ericsson, *Accounts for 2000*, Annual reports, 2001.

Appendix C
U.K. Telecoms Industry Statistics

Table C.1 shows the aggregate revenue and employment numbers recorded by the key players in the U.K. telecoms industry over the period of 1950 to 2000. They have been assembled from published sources, mainly company accounts. Where there have been gaps in the information, I have inserted my estimate of the numbers (shown in italics). Notes below describe in more detail the numbers reported. Revenue figures are shown in millions of pounds at their current price, and employee figures are shown in thousands.

The following comments relate to the main companies; see Section 2.3 for GPO/BT and Section 5.4 for other companies:

- *GPO/BT:* Numbers up to 1980 are derived from Parliamentary reports by the GPO. The numbers 1990 and 2000 come from BT accounts and include mobile phone (Cellnet) employees.

- *GEC:* The numbers from 1970 onwards include the acquisition of AEI, English Electric telecoms activities. The 1990 numbers include a 50% share of GPT (GEC Plessey Telecommunications) revenue. The numbers for 2000 include all GPT revenues.

- *Plessey/GPT:* The numbers from 1970 on include the acquisition of the Ericsson and ATM businesses. The 1990 numbers show GPT only, which was 50:50 split between GEC and Siemens. By 2000, GEC had bought out the Siemens stake.

- *Racal/Vodafone:* Telecoms revenue is solely that of Vodafone and its predecessor companies. Racal total sales numbers do not incorporate these

Table C.1
U.K. Telephone Industry Statistics

		1950	1960	1970	1980	1990	2000
GPO/BT	Telecoms revenue	100	242	652	3,601	12,315	18,715
	Telecoms employees	120	154	228	250	246	137
GEC	Total sales	55	117	891	3,006	8,786	7,625
	Telecoms revenue	10	20	100	150	909	1,858
Plessey/ GPT	Total sales	5	32	208	751	877	1,261
	Telecoms revenue	1	3	94	176	877	1,261
STC	Total sales	17	34	140	538	1,000	0
	Telecoms revenue	5	10	50	100	330	0
Racal/ Vodafone	Total sales	0	2	17	264	917	—
	Telecoms revenue	—	—	—	1	406	12,569
TMC/Pye	Total sales	1	25	50	—	—	—
	Telecoms revenue	1	3	5	—	—	—

Notes: Sales/revenue numbers are in millions of pounds at current levels. Employment numbers are in thousands.

revenue numbers but do include other telecoms activities (mainly export datacoms product sales of around £200 million in 1990). By 2000, the Racal parent company had divested its non-Vodafone assets primarily to Thomson CSF.

- *TMC/Pye:* The company was acquired by Pye in 1960. Total company sales show Pye for 1960 and 1967. In turn, Pye was acquired by Philips in 1966. Revenue numbers do not include the Philips group nor do telecoms include activities of the Philips group in the United Kingdom.

Appendix D
Telecoms Timeline

Table D.1

The Extended Timeline for Key Events in World History and the Telecoms Industry

Year	World/Corporate	Landline Telephony (Chapters 1 and 2)	Mobile Phone (Chapter 3)	Datacoms (Chapter 4)
1771	Start of the Industrial Revolution			
1783	End of the American Revolutionary War			
1792				First telegraph
1815	End of the Napoleonic Wars			
1838				Morse telegraph patent
1851				Telegraph between U.K. and France
1858				Ticker tape messaging
1861	Start of U.S. Civil War			U.S. telegraph coast to coast
1865	End of U.S. Civil War			
1872				Telegraph between U.K. and Australia
1876		First telephone		
1899		Strowger exchange		

Table D.1 (continued)

Year	World/Corporate	Landline Telephony (Chapters 1 and 2)	Mobile Phone (Chapter 3)	Datacoms (Chapter 4)
1918	End of World War I		German railway mobile	
1924				AT&T introduces fax
1926				Telex starts in Germany
1931				U.S. TWX
1933		U.K. BSA starts		
1938	PCM patent			
1945	End of World War II			
1947	Invention of the transistor			
1956			MTA Sweden	
1958		U.K. introduction of STD starts		
1961	First Post Office Act Plessey acquires ATE and BET			
1963		U.K. Highgate Wood Exchange		
1964				Xerox LDX patent
1965	Integrated circuit, Moore's Law		IMTS system U.S.	
1966	GEC acquires AEI	First PCM transmission		Packet switching developed
1967	GEC closes Woolwich			
1968	GEC acquires Marconi	Highgate Wood failure, U.K. introduces Crossbar		FCC Carterfone decision U.S.
1969	Second Post Office Act	End BSA		ARPANET USA
1971	Start of ICT wave			
1972		U.K. introduces TXE4 reed relay exchange		
1973			Cellular patent, U.S., first handheld phone U.S.	
1974				Ceefax U.K., TCP/IP

Table D.1 (continued)

Year	World/Corporate	Landline Telephony (Chapters 1 and 2)	Mobile Phone (Chapter 3)	Datacoms (Chapter 4)
1975		Fiber-optic transmission		
1976		System X order		X25 standard
1979	Election of U.K. Thatcher government			
1981	Privatization C&W	First System X delivery	NMT protocol, Nordic countries	
1982			U.K. licenses 2 network providers	Last U.K. telegram, French Minitel
1983			Cell phones introduced in U.S.	
1984	Privatization BT, breakup of AT&T		U.K. joins GSM discussions, first U.K. call	V32 modem U.S.
1985	GEC takeover of Plessey blocked	Last manual exchange U.K., BT orders AXE exchange, System X rollout begins in U.K.		
1986				
1987	GPT formed		EEC backs GSM	
1988		ISDN rollout in U.K.	Spinout of Vodafone	ISDN standard, Invention ADSL
1989	GEC/Siemens buys Plessey			World Wide Web
1990	Nortel acquires STC			
1992			Telepoint Europe	
1994				Netscape Navigator
1995	Last mechanical exchange U.K.			
1999	C&W U.K. operations sold			
2000	Dotcom crash, last reed exchange U.K.		U.K. auctions 3G bandwidth Japan i-mode	Peak sales fax machines
2001	AOL/Time Warner merger		First 3G network, Japan	

Table D.1   (continued)

Year	World/Corporate	Landline Telephony (Chapters 1 and 2)	Mobile Phone (Chapter 3)	Datacoms (Chapter 4)
2002	Collapse of Marconi/GEC			
2007	Bell reformed U.S.			
2008			4G rollout	BT stops Telex operations
2012				End of French Minitel
2018			5G starts	
2020	Huawei announcement			

Acronyms and Abbreviations

3GPP	Third Generation Partnership Project
ADSL	Asymmetric digital subscriber line
AEI	Associated Electrical Industries
AGSD	Advisory Group on Systems Development
AMPS	Advanced Mobile Phone System
AOL	America Online
APT	AT&T Philips Telecommunications
ARPA	Advanced Research Project Agency
ARPANET	Advanced Research Project Agency Network
ASCII	American Standard Code for Information Interchange
AT&E	Automatic Telephone & Equipment
AT&T	American Telephone & Telegraph
ATM	Automatic Telephone Manufacturing
BBC	British Broadcasting Corporation
BET	British Ericsson Telecoms
BICC	British Insulated Callender's Cables
BSA	Bulk Supply Agreement
BT	British Telecoms

BTS	British Telecommunications Systems Ltd.
CCITT	Consultative Committee for Telephony and Telegraphy
CDMA	Code division multiple access
CEPT	Conférence européenne des administrations des postes et des télécommunications (France)
CGE	Companie Générale d'Electricite (France)
CPRS	Central Policy Research Secretriat
CT2	Cordless telephone standard 2
CTNE	Compania Telefonica National de Espana SA (Spain)
C&W	Cable and Wireless
DBP	Deutsche Bundespost
DEC	Digital Equipment Corporation
DECT	Digital enhanced cordless telecommunications
DeTeWe	Deutsche Telephonwerke R. Stock & Co
DGT	Direction General (German) Telecommunications
DoCoMo	Do Communications over the Mobile Network
EDGE	Enhanced Data Rates for GSM Evolution
EE	Everything everywhere
EEC	European Economic Community
EV-DO	Evolution-Data Optimized
FCC	Federal Communications Commission (United States)
FDM	Frequency division multiplexing
FDMA	Frequency-division multiple access
FM	Frequency modulation
FSK	Frequency shift keying
FTTH	Fiber to the home
FTTK	Fiber to the kerb
GDP	Gross domestic product
GEC	General Electric Company (United Kingdom)

GPO	General Post Office
GPRS	General Packet Radio Service
GPT	GEC Plessey Telecoms
GSM	Groupe Speciale Mobile
HDML	Handheld Device Markup Language
HSPA	High Speed Packet Access
IBM	International Business Machines
ICL	International Computers Ltd.
ICT	Information & Communication Technology
IMTS	Improved Mobile Telephone Service
ISDN	Integrated Services Digital Network
ISP	Internet service provider
ITT	International Telephone and Telegraph
JTACS	Japanese Total Access Communications System
LAN	Local area networking
LG	Lucky Goldstar
LMT	Le Material Téléphonique (France)
LSE	London Stock Exchange
LTE	Long-Term Evolution
MCI	Microwave Communications, Inc.
MTA	Mobiltelefonisystem A (Swedish)
NEC	Nippon Electric Company
NEDO	National Economic Development Organisation (United Kingdom)
NHS	National Health Service (United Kingdom)
NMT	Nordic Mobile Telephone
NPL	National Physics Laboratory
NTL	National Transcommunications Limited
NTT	Nippon Telegraph and Telephone

NYNEX	New York/New England Exchange
OBR	Office for Budget Responsibility
OECD	Organisation for Economic Cooperation and Development
OFCOM	Office for Communications
OFTEL	Office for Telecoms
ONS	Office of National Statistics
PAM	Pulse amplitude modulation
PBX	Private branch exchange
PCB	Printed circuit boards
PCCW	Pacific Century Cyber Works
PCM	Pulse code modulation
PCN	Personal communications network
PSK	Phase shift keying
PTT	Posts, Telegraph, & Telecoms authority
QAM	Quadrature amplitude modulation
QPSK	Quadrature phase-shift keying
SBC	Southwestern Bell Corporation
SEL	Standard Elektric Lorenz (Germany)
SFR	Société Française du Radiotéléphone (France)
SIM	Subscriber Identifier Module
SITC	Standard International Trade Classification
SMS	Small Messaging Service
SPC	Stored program control
STC	Standard Telephones & Cables
STD	Subscriber Trunk Dialling
STL	Standard Telephone Laboratory
SWIFT	Society for Worldwide Interbank Financial Telecommunication
TACS	Total Access Communications System
TCP/IP	Transmission Control Protocol/Internet Protocol

TDM	Time division multiplexing
TDMA	Time division multiple access
TEMA	Telephone Equipment Manufacturers Association
TFP	Total factor productivity
TIW	Telesystem International Wireless
TMC	Telephone Manufacturing Company
TOR	Transmitted over radio (TALEX)
TUC	Trade Union Congress
TWX	Telewriter Writer Exchange Service
UNESCO	United Nations Educational, Scientific, and Cultural Organization
VAT	Value-added tax (U.K. sales tax)
WAN	Wide area networking
WCDMA	Wideband code-division multiple access
WiMAX	Worldwide Interoperability for Microwave Access

About the Author

John Polden lives in the United Kingdom and has a degree in electronic engineering from Southampton University. He obtained an MBA at London Business School and pursued a career managing U.K. electronics companies. He then went on to spend 20 years in the U.K. venture capital industry investing in technology start-up businesses. He spent 7 years researching the history of the U.K. electronics industry as an Honorary Research Fellow at the University of Sussex until 2023. Although he has contributed to several industry publications, this is his first published book.

Index

Artech House Mobile Communications Library

William Webb, Series Editor

For further information on these and other Artech House titles, including previously considered out-of-print books now available through our In-Print-Forever® (IPF®) program, contact:

Artech House	Artech House
685 Canton Street	16 Sussex Street
Norwood, MA 02062	London SW1V 4RW UK
Phone: 781-769-9750	Phone: +44 (0)20 7596-8750
Fax: 781-769-6334	Fax: +44 (0)20 7630-0166
e-mail: artech@artechhouse.com	e-mail: artech-uk@artechhouse.com

Find us on the World Wide Web at: www.artechhouse.com